NEUROPHYSIOLOGY

The Biological Basis of Mental Life

Also by Richard Stevko

ABSURDITIES -- b&w photos with annotations
ABSURDITIES II -- color photos with annotations
CONTRASTS -- postmodern poetry with guide
THE GHOST TREE - T'ang style poetry
    in collaboration with painter, Tamara Stevko
WHERE THOUGHTS COME FROM -- modern poetry
BEYOND A PERSIMMON -- metered poetry
MENTAL FUNCTIONING -- neuro-philosophic vocabulary
SHADES OF MEANING -- linguistic analysis of synonyms & antonyms
BEFORE PHILOSOPHY -- This exploration of the formation of philosophy

NEUROPHYSIOLOGY

The Biological Basis of Mental Life

R.M.Stevko, M.D

The Graven Image
Publishing

Hampden, MA

Copyright © 2014 by Richard Stevko

All rights reserved. This book or any portion thereof may not be reproduced or used in any manner whatsoever without the express written permission of the publisher except for the use of brief quotations in a book review or scholarly journal.

First Printing: 2014

ISBN 978-1-312-15196-3

The Graven Image, Publishing
Hampden, MA

Photo credits:
    Front Cover -
        stained pyramidal neurons in cerebral cortex.
        UC Regents Davis campus
    Back Cover -
        Raven and Richard
        courtesy Stevko archives

U.S. trade bookstores and wholesalers: Please email gravenimagepublishing.com or lulu.com

Dedication to: STUDENTS

TABLE OF CONTENTS

Introduction	ix
Part I foundational	
Divisions of the Nervous system	10
Microscopic	10
Neuron	10
Neurotransmitters	10
Macroscopic	12
central nervous system	12
brain	
spinal cord	
peripheral nervous system	12
spinal nerves	12
autonomic nervous system	13
cranial Nerves	13
Organization of the Nervous System	14
Sensory System	14
Central System	17
Encephalon	17
cerebrum	17
cerebellum	14
Diencephalon	18
Thalamus	18
hypothalamus	20
autonomic nervous system	21
limbic system	23
hippocampus	23
amygdala	23
Motor System	24
Cortical Mapping	25
Brain Centers	27
Part II clinical	29
Abnormalities of Motor Movement	30
Distinctions between Upper / Lower Motor Neuron	30
Cerebral Palsy	31
Seizures	32
Vascular	35
Syndromes and Conditions	36
Part III	43
Afterword - Selected Mental Functions	43
History of understanding Brain Function	54
Glossary, annotated	59
Bibliography	61

Preface

This book is an introduction to neurophysiology. It does not contain all that you want to know about the subject, but does contain all that you need to know to get started. It began as a handout for a graduate class of teachers and psychologists, who were interested not in pursuing advanced knowledge in neurophysiology, but felt the need for basic information in the field.

ACKNOWLEDGMENT

The author gratefully acknowledges the students who participated in the course after which this book is modeled. The largest group is the dedicated graduate teachers, who brought their expertise and dedication to the class conducted by one whose primary training is not as a teacher. The second group is the psychologists seeking advanced degrees who brought their clinical skills in a related area. To all these people who are not primarily biologists or related medically to the topic, I express my gratitude in helping me develop a practical approach to a topic as complex as the brain and as complicated as the mind.

Above all, Lorraine and Tamara, rigorously proofread, assiduously ferreted out embarrassing grammar and unclear descriptions. Their accomplishment is particularly valuable in that they, at the same time, absorbed the information and participated in discussions on the topics.

Tamara, particularly made available the skills of her design company, *Tamarartistry*, to make ready the diagrams and charts on complex points, rending invaluable assistance in designing clarification of muddles.

Introduction to Neurophysiology

The nervous system can be looked at in two ways, the way in which it is constructed (structure, anatomy), and the way in which it works (function, physiology). Since each of these perspectives contain enough knowledge to form a formidable discipline in its own right; but as both of these ways need to be understood together, by the novice, each will be superficially addressed and integrated without delay. Another reason for this holistic approach is that historically the anatomy greatly exceeded physiology in accumulated knowledge. Even today, the interaction of portions of the nervous system provide pragmatically superior explanations in the clinical environment.

Anatomically, the unit which serves communication depends on the order of magnitude of the question. If the order of magnitude is gross anatomy (the results of dissection visible to the naked eye) then the unit of communication is the nerve. If the order of magnitude of the question is microscopic anatomy (visible under the microscope) then the unit of communication is the nerve cell or neuron. If the order of magnitude of the question is molecular (chemical) then the unit of communication in the cell is electrolytic (Sodium and Potassium); or the unit of communication between cells is either neurotransmitters (across synapses), or hormonal (across stretches of circulation).

Physiologically, the nervous system is about communication. It is the vehicle which provides information from one area to another, mostly by electrical current along the nerve cells.

These answers telescope one within the other – the nerves are comprised of groups of neurons, the neurons are built of molecular ligands, which form the walls of the cell across which electrolytes pass or through which neurotransmitters or hormones are secreted.

In the opposite direction, macroscopically (to the unaided eye), neurons are grouped together into functional units. These units may perform a single function, discernible as a phenomenon; and, at the integrative end of the spinal cord (superior end) actually develop enough integrative connections to increase the volume of tissue and swell that area. Such ballooning, occurring repeatedly by evolutionary demands, has formed the parts of the brain.

Unfortunately for design purists, what we still call "systems" in the nervous system are not unitary systems, but cross integrated with other "systems", but still have enough vestigial function left to characterize them as centers for brain activity.
Beyond that, we realize that integrated systems really form interactive networks. The Afterwords gives a taste of that, but the topic is a different approach. To sample that, at each stage of this study, always ask, "What if this part connected everywhere else?"

PART I
Foundational

DIVISIONS OF THE NERVOUS SYSTEM

MICROSCOPIC

THE NEURONS are the building blocks of the nervous system. A single neuron consists of the cell body and extensions of that cell; namely the axon and dendrites, which are the extensions of the cell which allow it to contact other cells. A neuron usually has one axon, which is stouter that the dendrites and is efferent in function and transmits messages away from the cell; whereas the thinner multiple dendrites are afferent and transmit messages toward the cell.

Most cells in the body contain ions (electrically charged elements) of potassium ($K+$), and are bathed in body fluids rich in another ion, sodium ($Na+$). When the cell is stimulated by another cell, the sodium in the extra-cellular space exchanges place with the intracellular potassium, thus creating a transient difference in the trans-membrane voltages, creating an action potential, which is actually an electrical current flowing through the nerve cell. Information, which is transmitted from nerve cell to nerve cell, travels in the form of electrical impulses in the surface of the cell membrane. As the nerve cell membrane is stimulated, as the potassium exits from the cell, and the sodium enters, there is a change in the electrical charge upon the cell surface. When the action potential reaches the synapse, chemicals stored in bubble like structures are released into the synaptic space. These chemicals are called neurotransmitters and serve to stimulate (or inhibit) the next nerve cell in line to carry the message.

The dendrites of one cell "meet" the axons of another cell. The word *meet* is in quotes, because this meeting of the cell projections is not real, but virtual, as the cells always remain separated by a synapse, or space; which needs to be flooded by NEUROTRANSMITTERS, chemicals that modulate transmission of the electrical impulse.

Although there are over fifty neurotransmitters, and even more will surely be discovered, the following are so legendary that they are almost paradigmatic:

neurotransmitter	function	principle location
Acetylcholine	stimulatory	PNS (neuromuscular junction)
GABA	inhibitory	CNS
DOPA	helps signal transmission	CNS (hypothalamus)
Serotonin	regulates hedonic functions.	CNS & GI tract
norepinephrine	stimulatory	PNS (sympathetic system)

This is only a small sampling of the many neurotransmitters active in the brain. The locations listed for the neurotransmitters are not the only places they are found, but representative areas of heavy concentration. When the nerve impulse reaches the end of the dendrite, the vacuoles containing the neurotransmitter releases it into the synapse, allowing the chemical to flood the synaptic space, thus enhancing or delaying the passage of the nerve impulse; after which the chemical is reabsorbed into the presynaptic nerve. A large drug industry has grown around neurotransmitter types of drugs, drugs which mimic the neurotransmitters or drugs which enhance the timing of the drug in the synapse.

Most of the substances mentioned above are stimulatory in their functions. GABA is considered to be largely inhibitory. This has immense significance in the nervous system if we consider a series of cells most of which are stimulatory in sequence, but if we introduce one cell which is inhibitory. Consider if the cell preceding the inhibitory cell is stimulatory in nature; then the stimulatory cell is increasing the effectiveness of the inhibitory cell. The inhibitory cell is slowing the effectiveness of the stimulatory cells down the line from it. Consider the many ways in which these combinations can multiply the subtlety of the stimulatory – inhibition effect.

Further; the intensity of a single cell does not always depend on a feedback loop of cells to enhance its effectiveness, but may send a stimulatory dendrite back to its own cell body to stimulate itself, thus building up its own output. The reverse can also be true in the sense of inhibition.

In addition, most cells have a coating of Schwann cells to insulate them. These glial cells produce myelin which is the actual insulating substance. When this process of myelination is completed in the long tracts, the child can begin to walk. The other glial cells in the nervous system have functions which are not understood. They sometimes intrude their function when they develop tumors. Even at that time their function does not become apparent.

The cerebral cortex is composed of sheets of nerve cells. Microscopically, the nerve cells, called neurons, have a distinct cell body with projections called axons and dendrites. The axon (a cell body usually has only one) carries messages away from the cell body. The dendrites (a cell body usually has many) carry messages to the cell body. The space at which neurons contact each other to transmit messages is called the synapse. Macroscopically (unaided eye), in addition to neurons being organized into sheets, the cell bodies can aggregate into clumps (called nuclei within the Central Nervous System, and ganglia outside the Central Nervous System). Cell bodies, whether organized into sheets or clumps, form the gray matter of the brain. The pathways formed by axons and dendrites are called fibers, tracts, fascicles, and comprise the white matter of the brain.

Now that we have changed the order of magnitude of our perspective from micro to macroscopic, we shall discuss the topographic structure of the nervous system.

MACROSCOPIC - The nervous system is divided into the Central Nervous System (CNS) and the Peripheral Nervous System (PNS).

The central nervous system consists of the brain and the spinal cord.
The brain will be more appropriately discussed in the section on Organization.
The spinal cord descends through a bony protective layer called the spinal column, which is formed by bones called vertebrae stacked upon each other, bound by ligaments, and separated by gelatinous discs. The first seven vertebrae, and the segment of the spinal cord passing through it are referred to as the cervical spine; the next twelve, thoracic; five, lumbar; five, sacral; and the residual tail which is a fused mass of bone, (the coccyx). The spinal cord does not proceed into the coccygeal area.

The peripheral nervous system consists of the spinal nerves, the cranial nerves and the autonomic nervous system.

The spinal nerves project from the spinal cord through the intervertebral spaces to merge with other spinal nerves to form plexuses, and proceed to whichever part of the body they innervate. These nerves carry axons and dendrites, so there is not only information being transmitted from the body to the spinal cord (called afferent, or sensory fibers), but also information carried from the spinal cord to the body (called efferent, or motor fibers). The two types of fibers are indistinguishable in appearance.

The autonomic nervous system, sometimes called the visceral nervous system or involuntary nervous system, rises from nuclei in the brain stem and medulla and is closely connected to the hypothalamus, limbic system and endocrine systems.

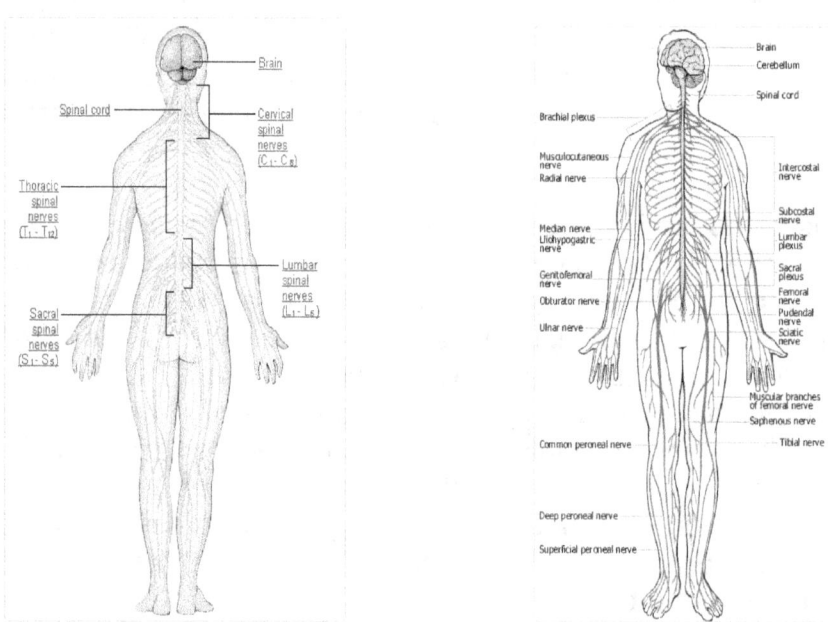

Fig. 1 = CNS & Spinal Nerves   Fig. 2 - CNS & Peripheral Nerves
Diagrams of the Human Nervous system, courtesy, Wikipedia

The peripheral nerves which project directly from the brain and exit the skull through holes called foramina and are referred to as Cranial Nerves.

Cranial Nerves There are many similarities between the cranial nerves and the spinal nerves. Both arise in clusters of nerve bodies (the cranial nerves arising in the nuclei of the forebrain, midbrain, and brain stem; the peripheral nerves arising in the anterior horn[1] of the spinal cord), both proceed through the substance of that portion of the central nervous system and exit in pairs to the exterior, and both (except in rare circumstances) contain motor and sensory fibers. The cranial nerves are twelve in number, and paired, just like the spinal nerves. Indeed, in the embryo, the nervous system develops from an invagination of the outer layer. The primitive neural tissue grows with the embryo and sends projections outwards to the rest of the developing body, projections, which become peripheral nerves. However, at the same time, the upper part of the early nervous system begins to balloon out into what will become brain. As a result, some of what would have become peripheral nerves are absorbed into the developing brain, and become cranial nerves.

The cranial nerves are referred to by name or by number:

I	Olfactory	smells
II	Optic	sees
III	Oculomotor	moves eyes
IV	Trochlear	moves eyes
V	Trigeminal	feels face
VI	Abducens	moves eyes
VII	Facial	moves face
VIII	Acoustic	hears, keeps balance
IX	Glossopharyngeal	swallows
X	Vagus	digestive functions
XI	Spinal Accessory	turns neck, shrugs shoulders
XII	Hypoglossal	tongue movement

---

[1] The names for the clusters of nerve cell bodies, nuclei and anterior horn, are an unfortunate reflection of their being named, by the ancient anatomists, according to appearance and not by function, which was not known at the time. In fact, the anterior horn is an extended nucleus progressing down the spinal cord.

ORGANIZATION OF THE NERVOUS SYSTEM

In keeping with the paradigm of nervous system organization, the sensory pathways bring information into the realm of the central nervous system, the brain processes the information, and the motor system carries the information out, resulting in an action or behavior. The fuel that drives the machine is sensory input. Experiments with sensory deprivation chambers showed the dramatic results, sometimes even ending in psychosis, when nearly total deprivation is prolonged.

SENSORY SYSTEM. The registration of information on a sensory nerve end is called sensation. The recognition, or interpretation, of that information in the central nervous system is called perception. Perception usually requires correlation of the sensation with information already in storage (memory).

People usually think of sensory information as being acquired through the "five senses". By that they usually mean sight, hearing, taste, smell, and touch. Those are the sensations received from the external world. There are also sensations received from the internal world, such as proprioception, balance, kinesthesia, or discomfort. It is difficult to know exactly where to stop this internal list, for several reasons. First, the internal stimuli are more vague than the external ones. For example, we can describe with reasonable certainty what is within our field of vision, or whether or not there is a stone in our shoes, but most people have difficulty telling the difference between a heart attack and indigestion. Secondly, most of us are not accustomed to discriminating between a pure internal sensation such as balance and a perception, such as keeping one's balance, which includes integrating sensory information with other information, either from memory or from other senses. Finally, there is no commonly accepted definition of where a simple perception stops and where our response to a perception begins. The distinction I am trying to draw here is between simple sensation (registration of information), perception (integration of that information with other information), and cognition or emotion. Biologically, the lines are not clearly drawn. Perhaps the issue will become clearer if we begin to describe the varieties of sensory systems.

Vision is effected by light rays entering the eyes and stimulating the lining on the back interior surface of the eyeball, called the retina. The retina contains naked nerve endings, which are stimulated, to varying degrees, by the light, which is focused on them. The stimulation of the nerve endings creates an action potential, which moves along the nerve fibers connected to the retina, resulting in a current of electricity moving along the optic nerve.

The Optic Nerve, remember, is a conglomeration of all the fibers connected to the retina in that eye. The information carried by that nerve is digitalized, that is, each fiber carries only the information that the nerve ending has or has not been stimulated. It is like a pointillist painting being cut up into its component dots, and each dot put into individual envelopes, and each envelope mailed separately to the same address, for reassembly at the destination back into a complete picture. The fibers, however, have a circuitous course from the eyes to the occipital lobe.

The course of the Optic Nerve, on its way to the back of the brain, involves the meeting of the two optic nerves in the midline of the skull, very soon after leaving the eye, where the fibers from the nasal side of each eye cross over to the other side (decussate), so the portion of the optic nerve, now called the optic tract, which proceeds beyond the decussation, contains fibers from the eye on the same side which originated in cells on the retina on the temporal side and also carries fibers which originate in the nasal side of the opposite eye.

These fibers synapse in a nucleus called the lateral geniculate. The synapses in the lateral geniculate are in precisely the same position, relative to all accompanying synapses, as are the retinal cells of origin. The postsynaptic fibers proceed to the occipital lobe, where they insert and register their packet of electronic information. The assembly of visual information in the occipital lobe represents a faithful mapping of the visual field, with the central part of the field being registered on the most posterior (farthest back) portion of the cortex, and is the sharpest. Information regarding location, distance, size, shape, movement and color are processed in an orderly fashion. As a result, apparently paradoxical phenomena, such as "blind sight" can occur, in which an isolated lesion might reduce the sharpness of an image to the extent that the person may be functionally blind, but can perceive motion, or brightness. Another strange phenomenon is achromatopsia, in which the person becomes blind only to color, but is otherwise perfectly sighted, seeing the world in black and white. These 'experiments of nature' underscore the wondrous way in which visual perception is reassembled within the head, instead of simply being projected onto a screen by an optic camera.
The sensation of vision is, at bottom, a phenomenon of transduction of light energy into electrical energy.

Hearing represents the transformation of sound waves to electrical activity, and the organ which accomplishes that is the cochlea. The cochlea works in conjunction with many portions of the hearing apparatus: pinna, ear canal, tympanic membrane, and ossicles.The pinna, which is commonly called 'the ear', acts as a reverse megaphone to channel sound waves down the external canal (which acts as a funnel for the sound waves), to the tympanic membrane, or eardrum, which separates the external ear apparatus from the middle ear.

At the tympanic membrane, the sound waves are converted into mechanical energy, which is modulated by the ossicles, three tiny bones (hammer, anvil, stirrup) that convert the movements of the eardrum into the dimension of wavelike signals to the fluid within the cochlea. Within the cochlea are nerve endings projecting into the fluid it contains. Therein, the waves stimulate the nerve endings, where the mechanical energy is converted to electrical signals. These signals are carried via the Auditory (eighth cranial nerve, VIII) Nerve to the Medulla and Pons, transmitted to the medial geniculate, then radiate to the Temporal Lobe, where they are interpreted as sound. VIII also carries signals from the vestibule, and that information is interpreted as the sensation of balance.

The Eighth Cranial Nerve has not been traced to the Temporal Lobe, but people with Temporal Lobe disorders do have communication disorders, so the association of a physical finding with consistent behavior in the same individual, repeated in all others with the same lesion, leads one to conclude that the information which started at the ear has to end up in the stated location. This concept will be returned to in the section on cortical functioning.

Taste is mediated by the taste buds on the tongue, which contain the nerve endings of Cranial Nerve VII on the anterior (front) two-thirds, and Cranial Nerve IX on the posterior (back) third. There are also many taste sensations from the pharynx and larynx. Individual taste buds are not very discreet at differentiating the tastes our perceptions can identify: sweet, sour, bitter, and salty.

Smell and Taste are closely related, as anyone with a 'cold' can attest. Most subtle tastes are probably a combination of smell and taste. The perception of taste is probably located in the parietal lobe, since gustatory hallucinations can be produced by stimulation of that lobe near the Rolandic fissure, and smell is probably mediated in the limbic system, considering the close association of smell with emotions. There is no cortical site at which stimulation can produce olfactory hallucinations, although they can occur with stimulation of parts of the olfactory bulb, and sometimes in the aura of temporal lobe seizures.

The close association of smell and emotion can be demonstrated easily by recalling how frequently smells elicit powerful memories, and how difficult it is to elicit a smell by any indirect means. Taste and Smell are extremely difficult to elicit as memories, unlike sight and sound, which are relatively easy. The transducers for taste and smell conversion to electrical transmission are chemical.

Touch receptors are widespread. There is no specific organ but global nerve endings associated with microscopic structures known as Paccinian Corpuscles and Meissner's corpuscles. Tactile senses may include pain and temperature.

The sensation caused by stimulation of these nerve endings are transmitted from the cutaneous surface, mucous membrane, or lining from which it arises and courses along the spinal nerve to the spinal cord at which point it enters the spinal cord through the posterior horn, and is transmitted up one of several spinothalamic pathways: the lateral spinothalamic tract for pain, posterior funiculus for touch and pressure, and the spinocerebellar tracts for stretch receptors.

Some definitions of touch sensations:
Anesthesia-loss of all form of sensation
Hypesthesia-reduced sensitivity to sensation
Hyperesthesia-increased sensitivity to sensation
Dysesthesia-painful sensation from an ordinarily innocuous stimulus.
Parasthesia- "pins and needles", burning, tingling feelings which arise spontaneously.

CENTRAL SYSTEM - The brain is located in the head and the spinal cord progresses down the back. The brain, from the top to the bottom, is composed of the encephalon (cerebrum), diencephalon (thalamus and hypothalamus), mesencephalon (midbrain), pons (bridge), and medulla (the last two of which connect to the cerebellum). The medulla progresses seamlessly into the spinal cord by leaving the skull through an opening called the foramen magnum.

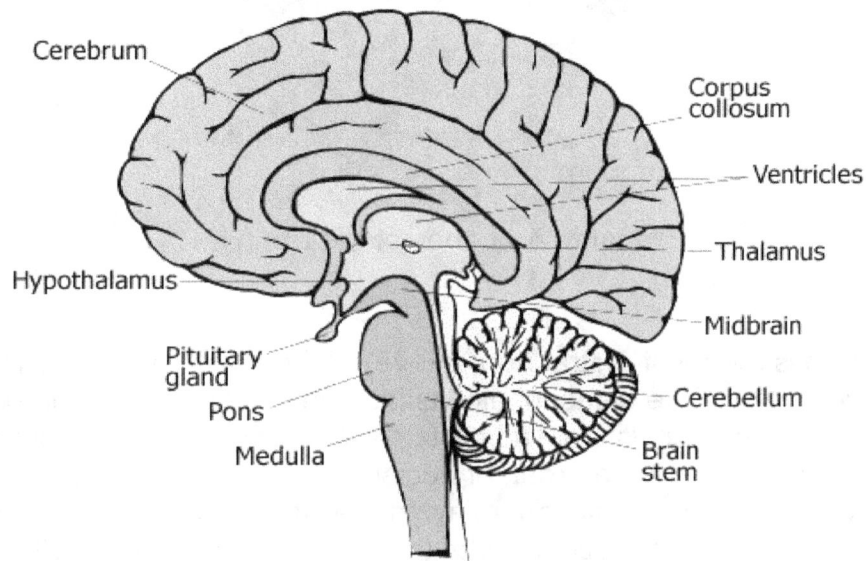

Figure 3 - Left lateral view of brain © The Epilepsy Association 2011

The Encephalon (cerebrum, cortex) is the most superficial part of the brain, forming its surface like a helmet, and forming hills (gyri) and valleys (sulci, fissures). The Rolandic and the Sylvian Fissures are critical structures for identifying the various parts of the cerebral cortex. The Central Sulcus of Rolando, is a vertical indentation in the upper half of the brain, about halfway between the front and the back of the organ, and progresses from the top to the second landmark, the Sylvian Fissure, which is a horizontal indentation proceeding from the front for about three quarters of the length of the brain surface (see Fig. 4).

The Rolandic Fissure separates the Frontal from the Parietal Lobes, and the Sylvian Fissure (lateral sulcus) separates the Parietal from the Temporal Lobes. The tip of the brain at the end of the Sylvian Fissure, connecting the Parietal and Temporal Lobes is the Occipital Lobe. The cerebral cortex is divided into two distinct hemispheres, left and right, being connected by communicating nerve tracts called the corpus callosum.

The cerebellum lies tucked under the back of the brain and overlies the medulla and pons, to both of which it is connected; placing it in prime position for fine tuning motor activity and controlling motor memory.

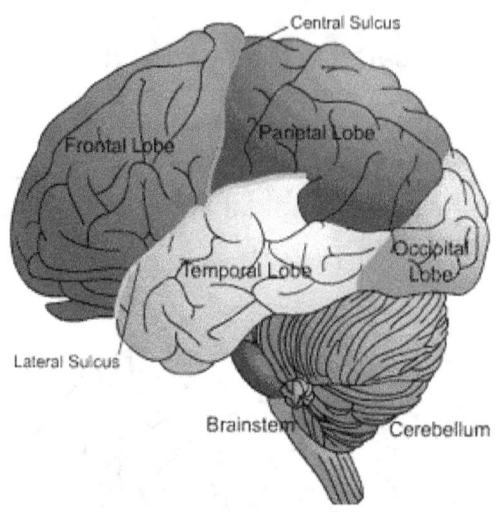

Figure 4 - Lobes of the Brain

Courtesy of WordPress software

The Diencephalon is the most rostral (top, above) of the major structures in the brain, except for the cerebrum. The diencephalon is located just below the cortex and above the mesencephalon, and consists of two parts: Thalamus and Hypothalamus (hypo=below). The thalamus is a group of nuclei which act in a unified way, the aggregate performing a single function, relaying messages from lower centers to the cortex or other thalamic nuclei. It consists of the *thalamus* and the *hypothalamus*. The similarity in their names is deceptive and arises from their anatomic proximity, rather than similarity of function. The word *thalamus* comes from the Greek word indicating an inner chamber of the house, and hypothalamus is under the thalamus.

The Thalamus is actually a series of nuclei which act as final relay stations for messages going to the cerebral cortex, and seem to serve some kind of distribution control function.

In this crucial location, the thalamus is well located to perform its primary function, relaying information from below. The relay function of the thalamic nuclei serves several purposes: a)sorting, b) disseminating.Examples of the sorting process can be listed in this enumeration of thalamic nuclei which sort incoming information and distribute it to higher centers:

nucleus	connects
lateral geniculate	optic tract to occipital visual cortex
medial geniculate	auditory nerve to temporal lobe
posterior ventral nucleus	sensory input to somesthetic parietal cortex
anterior nucleus	hypothalamus to limbic system
dorsomedial nucleus	frontal lobe cortex to hypothalamus
anterior ventral nucleus	corpus striatum to frontal motor cortex
ventral intermediate n.	cerebellum to prefrontal motor areas
lateral nucleus	interconnects parietal association areas
pulvinar	patietal-temporal interconnections
habenula	olefacory relay

As we walk down a hallway, or run through the woods, or drive in traffic, or swim in a pond, or make our way through a dream, or suffer a bellyache, we are surrounded with a morass of data, external and internal, which comes to us through the sensory system. A portion of it reaches awareness, and the rest is processed by the central nervous system without conscious effort. The process of sorting that data is a function of the thalamus.

The thalamus can be thought of as receiving the masses of sensory data coming into the central nervous system and sorting them to appropriate places for attention, thereby serving the functions of correlation and integration (making sense of) sensation on the way to becoming perception.

The second primary function of the thalamus is to create a partial awareness of peripheral sensory stimuli. If we think of "having an idea" as being a full cortical function, we can think of the thalamic portion of that as a preliminary mental step, namely focusing, or becoming aware of, the necessity to formulate thought regarding a particular sensation we receive.

The third function of the thalamus is to focus attention--temporarily making certain cortical areas more sensitive and receptive to the incoming data.
These functions seem paradoxical in that it seems the thalamus is making decisions regarding the priority to which data will be assigned, while we have assigned the executive functions of deciding and assigning to the cortex. This paradox is not all that unusual in the sense that we cannot ever know anything new without some prior knowledge about it. In the days of naivete, when there was something new under the sun, explorers would return to civilization with tales of sea monsters, gorillas, centaurs, and other wondrous sights, sounds, and other sensations. The public, having no knowledge of these things would learn of them by comparing descriptions, in their mind's eye (i.e. occipital cortical association areas), and formulating a concept to which additional information could be added as it became available.

In the same way, we learned, as innocent children to take data, compare it with storage in memory, and build concepts which became future memories upon which further knowledge added or subtracted information to make the never ending process of learning come closer and closer to reality. As innocence passed and sophistication settled in, we further expanded that process to build metaphoric analogies, many of which became realities in themselves forming an entire complex of ideas so palpable we accepted them as dogma. These truths form a large part of our mental lives, and the life-long process of continually discriminating between what is real and what is fantasy is the very real process of maintaining sanity. Conversely, the opposite process of being able to move comfortably between reality and fiction maintains creativity.

Also, just as learning is enhanced by an interactive Socratic give and take, fibers pass from the thalamus to the cortex, and fibers return, establishing a reciprocal effect, an interactive dialogue.

The Hypothalamus - has both endocrine and non-endocrine functions. It also controls appetite and body temperature, participates in emotions (through it's connection with the limbic system) and wakefulness (through it's connections with the Reticular Activating System).

The endocrine functions have to do with hormone secretions, which are expressed primarily through a gland (which projects from the under surface of the hypothalamus) called the pituitary gland. The pituitary is also known as the 'master gland' because it's secretions or hormones travel through the blood stream to control all other endocrine glands in the body. Endocrine glands differ from the other glands in the body, which are called exocrine glands, in that they secrete into the blood stream, and influence other cells by the influence of these hormones being carried throughout the body. Exocrine glands, on the other hand, secrete into a cavity, or space, and influence the substance found there. The exocrine glands are not under hypothalamic control.

Some examples of glands are:
<u>Exocrine</u>
digestive glands, pancreas-secrete enzymes into the digestive cavities which break down food.
sweat glands-secrete sweat onto the skin for thermal control.
mammary glands-secrete milk from the breast for nourishment of the young.
<u>Endocrine</u>
thyroid-controls metabolic activity of all cells.
pancreas-controls sugar metabolism through the hormone, insulin.
adrenals-secrete hormones to control sugar, mineral, steroid balance; and adrenaline.
gonads-secrete sex hormones.

An important aspect of endocrine control is the *feedback circuit*. When any endocrine gland, for example, the thyroid, secretes an amount of thyroid hormone(s) into the blood stream which is inadequate, the decreased level in the blood, which also washes the pituitary, is detected as being 'inadequate' by the pituitary and the pituitary is induced to secrete additional amounts of thyroid stimulating hormone (TSH) into the blood, which directs the thyroid to make more thyroid hormone. When that happens, the level of thyroid hormone is increased in the blood stream, and the pituitary, sensing this increase, secretes less TSH, resulting in relaxation of the thyroid's efforts, until the levels of thyroid hormone fall to inadequate levels again; then the cycle repeats itself. This is much like the thermostat in a building. When the furnace (thyroid) produces inadequate heat (thyroid hormone), the thermostat (pituitary) detects the air cooling (decreased levels of thyroid hormone in the blood stream) and responds (TSH) by directing the furnace to make more heat. When the air is warm enough (adequate levels of thyroid hormone), the thermostat shuts off (the pituitary decreases TSH secretion) until there is balance.
Similar mechanisms are employed by the pituitary to control the other endocrine glands in the body and these are described in any elementary textbook of human physiology.

The hand that sets the thermostat is the hypothalamus. The pituitary gland is connected to the hypothalamus by a narrow stalk, through which nerve fibers, and around which a plexus of blood vessels, run. Hypothalamic control of the pituitary is by neural messages and by neuropeptides elaborated in the hypothalamus, itself. These neuropeptides act like neurotransmitters, in that they are manufactured by the nerve cell, and they act like a hormone, in that they influence other cells after being carried to them by a closed circulatory system. They are unlike neurotransmitters in that the influenced cell is not another neuron, and they are unlike hormones in that they are not deposited in general circulation. This exciting field of neuroendocrinology is rapidly expanding. Some of these "hormones" are:

    Thyrotropin-releasing factor (TRH)
    Growth hormone-releasing factor (GHRF)
    Corticotropin -releasing factor (CRF)
    Gonadotropin-releasing hormone (GnRH
    Prolactin-releasing factor (PRF).

These secretions find their way to the anterior pituitary (the front part of the pituitary gland) also called adenohypophysis. Hypophysis is a synonym for pituitary. The nerve fibers go to the posterior (back part) pituitary, also known as the neurohypophysis. In the adenohypophysis, these factors influence the production of Thyrotropin (TSH), Adrenocorticotropin (ACTH), Leutenizing Hormone (LH), Follicle-stimulating Hormone (FSH), Growth Hormone (GH), and Prolactin.

The Neurohypophysis produces Vasopressin and Oxytocin. Vasopressin has an antidiuretic effect on the kidneys, and oxytocin initiates uterine contractions as well as milk ejection.

As can be seen, the hypothalamic-pituitary axis is exceedingly complex and elegant; it is in turn influenced by many non-endocrine factors in the body.

The non-endocrine functions of the hypothalamus include the **autonomic** nervous system. The autonomic nervous system is an involuntary system which also operates on a feedback basis. As a need is perceived, the hypothalamus can respond in one of two ways (increase/decrease) very much as a thermostat does in a house to increase or decrease heat. The response of the autonomic nervous system is more complex than a thermostat in that it also controls the circulatory system, the respiratory system and the digestive system. This control is exerted through two sets of wiring called the *sympathetic nervous system* and the *parasympathetic nervous system.* Both systems are always turned on, and the net effect is the result of which is turned higher, very much like filling a bathtub with hot and cold water in such balance that the resultant bath is titrated to the soaking taste of the person whose hands are on the faucet. The sympathetic division has a thoracolumbar outflow, and the parasympathetic division has a cranio-sacral outflow. These designations simply indicate the point of exit from the central nervous system to the periphery. Once in the periphery, the autonomic nervous system forms a relay in clusters of cells called ganglia. Both divisions end innervating the same organs.

The sympathetic system, when stimulated, results in an increase in the rate (cardiotachic) and the strength (cardiotonic) of the heart beat. This modulation of the cardiac pump is an elementary mechanism to increase circulation. The blood vessels constrict, raising the blood pressure, and the bronchial tubes relax, allowing for increased air flow to the lungs, resulting in greater oxygenation of the increased circulation. This result, when physical activity is called for, or in states of excitement is referred to as the *fight or flight* reaction. Other effects are summarized in the table.

The parasympathetic system, on the other hand, causes the reverse effects. The heart rate is slowed, the blood pressure drops, respiratory rate slows and bronchial tubes constrict. The digestive system increases both secretions of digestive enzymes and rate of bowel activity. This effect is referred to as the *rest and relaxation* response.

AUTONOMIC NERVOUS SYSTEM

Parameter	Sympathetic	Parasympathetic
Heart Rate & Strength	increased	decreased
Blood Vessels, Peripheral	constricted	relaxed
Blood Vessels, Cardiac	relaxed	constricted
Bronchial Tubes	relaxed	constricted
Digestive Functions	relaxed	increased
Pupils	dilated	constricted
Sweat	increased	decreased
Piloerection	increased	decreased

It is easy to see how the hypothalamus participates in body temperature control through it's influence on the autonomic nervous system, with the ability to manipulate blood flow and vascular dilation resulting in heat loss or conservation. Through the same system it participates in the expression of emotions. In fact, physiologic psychologists, for a long time, considered the autonomic response to be the motor expression of emotions. This is also easy to understand if we think of times of excitement (heart pounding i.e., cardiotachic and cardiotonic effects) and times of embarrassment (flushing). The hypothalamus does participate in the expression of emotions through the autonomic nervous system, but it also participates in ladening perceptions with emotional meaning (see the Limbic System).

The hypothalamus is involved in appetite, although nutrient as well as sexual seeking activities have many other parameters associated with them: physiological, social and emotional. Nevertheless, lesion analysis in humans and direct stimulation studies in animals suggests a strong influence of ventromedial nuclei in satiety, and lateral nuclei in appetite. See also: Froelich Syndrome, Diencephalic Syndrome, Prader-Willi Syndrome, and Klein-Levin Syndrome.

Wakefulness is also attributed to the hypothalamus through it's connections with the Reticular Activating System (RAS). The RAS is a loose formation of neurons extending from the brain stem to the posterior diencephalon. Lesions in this area reduce awakedness, and this area is richly innervated with many sensory neurons. It is entirely likely, but not yet demonstrated, that the RAS acting through the hypothalamus uses it's connections with the limbic system to enhance learning and memory.

The Limbic System is not a structure, but an interaction, whose main components seem to include the hypothalamus, hippocampus, and amygdala. These, and surrounding structures, have rich and complex inputs from the sensory system, the RAS, and the cerebral cortex, primarily; and output messages to the cortex (for storage in the association areas, as well as stimulation of the motor system), the endocrine system, and the autonomic nervous system. Its functions involve emotion, memory, and what collectively can be called survival strategies (behaviors surrounding feeding, reproduction, nurturance, etc.)

Our knowledge of the components of this interactive group of structures, this system are summarized:

HIPPOCAMPUS

Location - The hippocampi (pl.) are bilateral seahorse shaped structures located in the medial temporal lobe, underneath the cortical surface.

Functions include:
- memory -
  - almost universal agreement that important role in memory
  - the precise nature of this role remains widely debated
    - consolidation of information
    - (short-term memory to long-term memory
- spatial navigation -
  - theories about "cognitive maps" in humans and animals.
  - as with the memory theory, almost universal agreement
    - spatial coding plays important role in hippocampal function,
    - but the details are widely debated.
- inhibition - least popular of the three functions
  - based on hyperactivity (inhibition release) in hippocampal damage
  - demonstration of increased hippocampal size in ADHD

AMYGDALA

Location - The amygdalae (pl.) are almond-shaped groups of nuclei located within the temporal lobes of the brain in complex vertebrates.

Function - They perform a primary role in the processing of memory and emotional reactions. Specifically, More recent studies [LeDoux] indicate that (A) A unique (because specific in this area of nebulous emotions) act of the amygdala reproduces one of its distribution functions, where incoming information is not only sent directly to the cortex, but also short circuits the information to the autonomic nervous system, so the information (deemed threatening) causes a visceral reaction ("freezing", "fight or flight") even before the information is cognized for rational action. (B) fuller understanding of the amygdala is gained from its widespread, again, nebulous interaction with surrounding structures in the juncture of the temporal and parietal lobes (TPJ) and the interfaces (Insula) of those lobes as they approach the frontal lobe.

It has been noted by many theorists of brain function that the intermediate position held by the limbic system between cortical outflow and the hypothalamus is representative of the way perceptions (a cortical function) is mediated into visceral responses (hypothalamic effects), and puts the feeling into emotions, memories, and instills learning with meaning.

Motor System Earlier, it was pointed out that the simplified version of neural functioning consists of input (sensory system), modification (CNS), and output (motor system). If we can conceive of a simple reflex arc in which sensory afferents synapse in the spinal column with motor efferents, we would have reproduced an actual reflex in which no modifying factors have intervened. Such reflexes occur when one burns a finger and pulls away without thinking about it. Similarly, when the patellar tendon is percussed with a reflex hammer, the knee jerks.

Both of those examples seem instantaneous and unconsidered. However, there are many layers of controls, which are involuntary. The known levels, which modify both volitional and seemingly involuntary movement, include:
- the motor cortex
- the brainstem nuclei
- the subcortical systems
- the premotor and accessory motor cortices
- the sensory systems.

THE MOTOR CORTEX is the part of the frontal lobe just in front of the Rolandic sulcus, and contains peculiarly large cells (Betz cells) which are directly connected to the spinal nerves in such a way that stimulation of the Betz cells causes movement of a body part so consistently that a map could be drawn onto the motor cortex to show to which parts of the body that part of the cortex is connected. This representational map is called a homunculus. The pathway formed by the axons of the Betz cells in the motor cortex as they descend to the part of the spine at which they form their synapse is known as:

THE CORTICOSPINAL TRACT (also known as the pyramidal tract, from its most apparent portion, which resembles a pyramid; also known as the upper motor neuron, to distinguish it from the peripheral nerve, which is called the lower motor neuron. It will become apparent in the next section why this distinction has persisted).

THE BRAINSTEM NUCLEI are clusters of nerve cell bodies lying in the brainstem, and influencing the muscle group activities that have to do with posture and with repetitive, automatic movements.

THE SUBCORTICAL SYSTEMS include the basal ganglia and the cerebellum, which modify tone, posture, and coordination.

THE PREMOTOR AND ACCESSORY MOTOR CORTICES are the parts of the cerebral cortex, which activate before the motor cortex is stimulated, and are thought to be part of the planning and programming activities, which the motor cortex carries out.

THE SENSORY SYSTEMS provide the information upon which all the other parts of the motor system act, e.g., visual information is brought into the central nervous system and used by the premotor and accessory motor cortices to influence motor movement, such as not walking into a wall. In the same way the information from the other senses informs the decisions carried out by the motor system. The sensory information includes not only information from the traditional five senses, but also information transmitted by the sensory system from the interior of the body. In a broad sense, then, all the information we get from sight, hearing, smell, touch, and taste, as well as all the information we get from discomfort, position sense, inadequate hormone secretions, and hunger are sensory information and transmitted by afferent nerves to the central nervous system. Conversely, all the movements we make, voluntary and involuntary, including stepping aside from danger, as well as each heartbeat and all peristaltic contractions of the intestines are motor activities.

In a more limited way, the voluntary muscular movements are involving signal transmission from the Betz cells down the corticospinal tract to synapse with the beginning of the lower motor neuron in the anterior horn of the spine. This part of voluntary motor movement is called the pyramidal tract. All other influences on the pyramidal tract, such as the above described brain stem nuclei, cerebellum, and basal ganglia are referred to as **extrapyramidal** tract influences. We shall further discuss the clinical implications in Part II.

CORTICAL CENTERS

Cortical mapping the result of cytoarchitectonics (a variation of the term cytoarchitecture) meaning a study of the function of different portions of brain tissue and/or the structure of the cells which comprise those portions under study. In the late 1800s, brain function was still primarily understood as the result of specific centers in the brain being responsible for certain actions. Phrenology [GALL] represented the extreme of this kind of thinking, and though discredited on its face played an important part in the emergence of a naturalistic approach to the study of man including development of evolutionist theories, anthropology, sociology and establishing psychology as a science.

When (19th cent) it was noticed [MEYNERT] that the cell structure of various parts of the grey matter had a lot of variation. Several neuroanatomists immediately, tried to divide the cortex into homogenous parts, based on function, and/or structure. The most widely used classification in clinical neurology [BRODMANN] divided the human cortex into 44 areas.

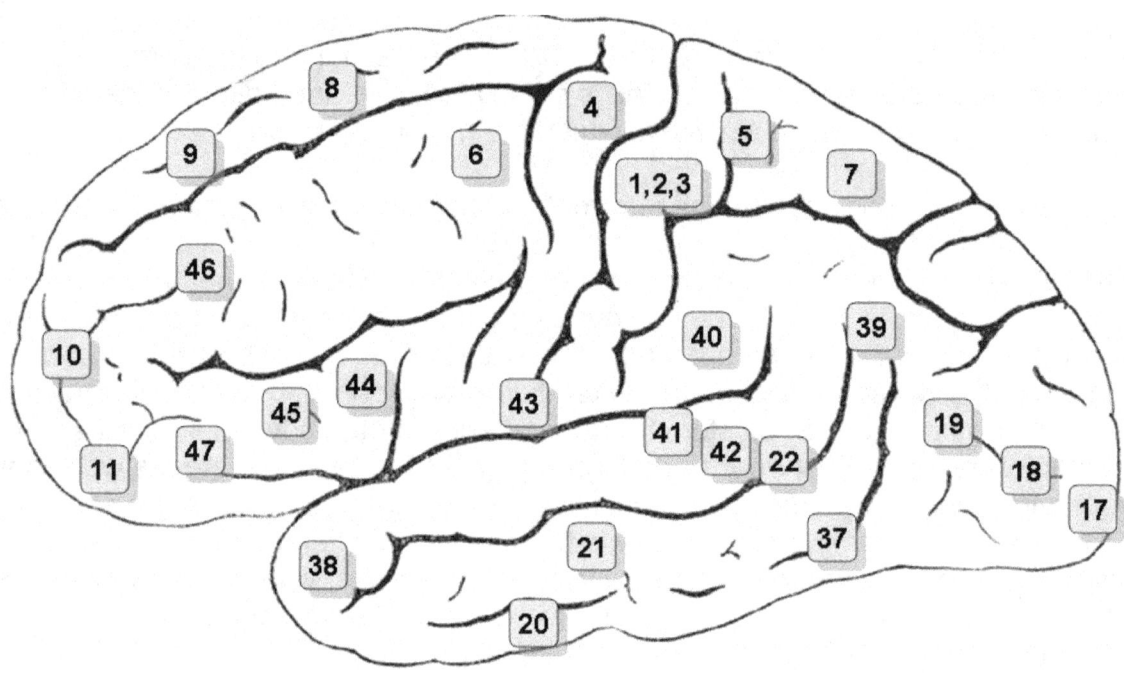

Fig.5 Brodmann's areas on Lateral surface of left cerebral hemisphere
from *20th U.S. edition of* Gray's Anatomy of the Human Body

Only a few of these areas have been demonstrated to have dedicated functions (see History - Part III) and are designated as Primary Cortices:
Brodmann's area number 17 (primary cortex) is directly stimulated by the eye and in turn stimulates areas 18,19, (association cortices[2]) forming the visual system;which in turn stimulates, or is stimulated by other areas - creating perception. In the same way, are 41 receives sensation from the ear, and refines hearing in areas 42, etc and is perceived by interaction in other association areas. Primary area 1-3 is receptor for sensation and passes the information to other association areas forming related perceptions. Obversely 4-6 is the origin of motor activity, having been stimulated by premotor areas which are not primary areas, but influenced by the pre-motor areas, being, in an existential sense, the end of the road for perception, the channel for action.

---

[2] unimodal or heteromodal, depending on specificity.

A unique categorization [MESULAM] of cortical organization connects the idea of five basic anatomic types with five degrees of cortical specificity:

	EXTRA-PERSONAL SPACE	
Idiotypic cortex	Primary Sensory and Motor Areas	the first cortical relay from sensory organs (motor from premotor area)
Homo-typical Isocortex	Unimodal Association Areas	stimulated by primary sensory cortex
	Heteromodal Association Areas	stimulated by unimodal or heteromodal cortex
Paralimbic Areas		stimulated by heteromodal and few "downstream" unimodal cortices
Limbic Areas		interaction cortex / hypothalamus
Hypothalamus	INTERNAL MILIEU	

In the range of cerebral zones from Idiotypic to Limbic the cellular architecture goes from well differentiated to simple, the evolutionary stages are from most recent to most primitive, and the specificity of purpose goes from most dedicated to most general.

The other areas, as they approach the functions of the internal milieu, are much less specific in function, e.g., internal discomfort is less easily described than surface disorders, and states like emotions are more nebulous than physical activity. As we leave the cortex to speak of the autonomic areas, as we did, we see that many of those functions are involuntary and difficult to subject to volitional control.

Brain Centers -- We can't talk about brain mapping without clarifying the issue of "centers". This may be a function of this time in history, but not a month goes by without a news report arising regarding the "discovery" of a brain center for appetite, pleasure, language, which is promised to solve related problems of obesity, unhappiness, or shortcuts to communication. Very few of these are dedicated to only the item of interest, but are really areas that cover many aspects of the topic.

The idiotypic cortex for sensation (S1) and motor stimulation (M1) is almost exclusively dedicated to feeling and moving. The visual cortex (V1) and the auditory cortex (A1) are likewise highly specific, being the first cortical relay from the sense organs.

Other so called specific centers are really unimodal association areas with links to idiotypic cotices and multiple links to other association areas. Areas like the speech centers of Broca's area (Brodmann's areas 44, 45) and Wernicke's area (Brodmann's area 22), require other associations for proper functioning. Wernicke's, for example requires that most of the core difficulties come from damage to the medial temporal lobe and underlying white matter. [Kolb]

Broca's aphasia is the loss of the ability to produce spoken or written language (expressive aphasia). Comprehension is preserved, but the speech is at best agrammatic.

Wernicke's aphasia is a difficulty in receiving information by spoken or written language (receptive aphasia); though they can speak with normal grammar, syntax, rate, and intonation, they cannot express themselves meaningfully using language.

PART II
Clinical Examples
of
Neurological Dysfunctions

Clinical examples of Neurological Disorders are added to or integrated with the course this book served. They were integrated as performance pieces to clarify the principles discussed. They remain as supplemental material because, in a sense, even the pioneering scholars from ancient to modern times have learned from dysfunctions.

Most of these examples are associated with children because most of the students are teachers, and because that is the author's area of expertise. Strokes (brain attacks) require special mention here because the damage done to the brain when the blood supply to that part is blocked has provided investigators, who were usually treating physicians, with information about that part's function.
The seemingly arbitrary designations differentiating upper motor neuron/lower motor neuron, and pyramidal/extrapyramidal persist in our language because they reveal important distinctions in disease states.

ABNORMALITIES OF MOTOR MOVEMENT

Upper and lower motor neuron distinctions become apparent when we realize that damage to either part results in paralysis, but the results from the location of the damage cause different types of paralysis, and is of prime importance in determining the diagnosis of the cause of the paralysis. The upper motor neuron is the part of the motor system, which lies between the motor cortex and the anterior horn cell; the lower motor neuron lies between the anterior horn cell and the neuromuscular junction. In simplistic terms the synapse preceding the anterior horn cell divides the upper from the lower motor neuron.

Any damage to the upper motor neuron, such as lack of oxygen to the cortex or a blow to the head or a tumor impinging upon the corticospinal tract results in paralysis or weakness (paresis) in the part of the muscular system provided by the damaged part, and the location of the paralysis will be necessarily indistinct. Additionally, the provided muscles will develop increased tone. Tone is resistance to passive stretch. As a result, there will also be increased resistance to passive movement (increased reflex). This combination of hypertonia and hyperreflexia is called spasticity. The atrophy (wasting) which results is proportional to the limitation of use caused by the paralysis or paresis, and is referred to as disuse atrophy. This process results in a number of pathologic reflexes, the most notable of which is the Babinski reflex, in which scratching the sole of the foot causes the great toe to rise, instead of go down, and the remaining four toes splay, instead of curl.

Damage to the lower motor neuron, such as injury to the nerve beyond the beginning of the anterior horn cell, will result in paralysis of only the muscle innervated by the damaged nerve, decreased tone and reflexes (flaccidity), and profound atrophy. The atrophy in lower motor neuron injuries often involves up to 80% of the affected muscle mass, and is not remediated by exercise. In the process of atrophying, the individual muscle fibers go into contractions, causing a peculiar twitching appearance on the surface known as fasciculations.

	upper	lower
site of paresis	**indistinct**	**limited**
tone	**spastic**	**flaccid**
atrophy	**disuse**	**profound**
other	**pathologic reflexes**	**fasciculations**

Pyramidal and Extrapyramidal system distinctions are based on the fact that the pyramidal tract represents a straight connection from the motor cortex to the lower motor neuron. The extrapyramidal system includes all the modulating systems: basal ganglia, cerebellum, and brain stem nuclei.

Both systems are upper motor neurons, and, as such, one component of damage is spasticity. Spasticity in the pyramidal tract is usually shown as **clasp knife rigidity**, and is present in the flexors of the arms and extensors of the legs. In the extrapyramidal tract, it shows as either **lead pipe rigidity**, or **cogwheel rigidity**, and is present in the extensors of all limbs. Damage to the pyramidal tract results in paralysis (loss of movement) or paresis (weakness), depending on how many fibers are destroyed. Damage to the extrapyramidal system may cause loss of movement, depending on the extent of damage, but is more characterized by involuntary abnormal movement.

The types of abnormal movement can be:
- Chorea
- Athetosis
- Dystonia,

or any combination of the above.

Cerebral palsy is usually the result of damage to a combination of multiple sites in the extrapyramidal tract, and results in variable combinations of abnormal involuntary movements (A.I.M.s). It is so named because the original condition was understood to be a palsy (paralysis) resulting from damage to the cerebral cortex. The name of the condition is a descriptive term and is not a diagnosis. Most of the damage is in the pyramidal and extrapyramidal tracts, and not always in the cerebrum. The abnormal movements are usually chorea and athetosis, rather than paralysis.

Paralysis-complete inability to move.

Paresis-weakness of muscle groups, which if the etiologic lesion were more extensive would result in paralysis. Paralysis and Paresis are relative terms for degrees of inability, and can be caused by disorders, damage, or disease of the pyramidal tract (cortical motor neurons, corticospinal tract) or peripheral nerves. Problems in the extrapyramidal tract usually result in:

Chorea-uncontrolled movements which are sudden, jerky, and uncontrolled. The word comes from the Greek verb *to dance*. A common finding in cerebral palsy, as well as in Sydenham's chorea (also known as St.Vitus' Dance, a complication of streptococcal infection, and previously commonly seen in connection with Acute Rheumatic Fever), and Huntington's Chorea (a hereditary disorder which has it's onset in the fifth decade of life.

Athetosis-slow, sinuous, writhing movements, which are also involuntary. The word also derives from a Greek word meaning *unfixed*. Also commonly found in cerebral palsy, where a combination of the two abnormalities are found together, and referred to as choreo-athetosis.

Ballismus-a sudden, uncontrolled movement which is identical to the action taken in throwing something. The word has the same root as *ballistic*.

Tremor-uncontrolled oscillating, to and fro movements. Often found in Parkinson's Disease, and as a complication of antipsychotic drug use.

Dyskinesia- is a general term for abnormal (dys) movement (kinesia). All of the above conditions qualify as dyskinesias.

Tardive Dyskinesia or T.D.-a combination of, in several forms, of facial twitches or grimaces, pill rolling tremors, rigid posturings, which have resulted from long term antipsychotic medicine usage.

Akathesia- a motor restlessness which may be manifest by marching movements when not walking and sometimes while sitting, or stamping of the feet, or other restless uncontrolled movements.

Tic-an involuntary motor movement which is repetitive and stereotypic. The cause is no known.

Seizures-uncontrolled motor movements caused by abnormal discharges of electrical activity on the cerebral cortex.

Epilepsy **SEIZURES** (convulsions, fits) are intermittent events in which there is an a "sudden, excessive, disorderly discharge of cerebral neurons"-Hughlings Jackson. The reason for this discharge may be an irritated area or damaged area, an inherited propensity, or metabolic changes. Having repeated seizures is called epilepsy. The classification of seizures proposed in 1981 by the International League Against Epilepsy, and widely accepted by neurologists, is based upon 1) the extensiveness of the seizure, and 2) the degree of consciousness alteration.
Partial Seizures are limited to a part of the body and reflect a highly localized locus of activity in the cortex. They are further subdivided into *simple* (no loss of consciousness) and *complex* (consciousness is lost). A third type of partial seizure is one in which the

motor activity begins in a part of the body then is *secondarily generalized* to the rest of the body.

Simple Partial Seizures are sometimes called simple motor seizures or Jacksonian seizures. Clinically they differ, but the significance in localizing the irritated focus is the same. Classically, the simple motor seizure consists of a sudden turning of the head and eyes, often accompanied by tonic contraction of the trunk and extremities on the side from which the head is turned away. This particular form is often interpreted by observers as a volitional event, or involuntary 'behavior', which some therapists unsuccessfully try to 'modify'. The Jacksonian seizure (named after Hughlings Jackson, who first described the phenomenon) consists of discreet tonic movements in the fingers of one hand, on one side of the face, or one foot, then converts to a clonic movement which then "marches" (spreads) from the affected part to other parts on the same side of the body. If either of these simple partial seizures becomes generalized (see below), the person loses consciousness.

Complex Partial Seizures are also called *psychomotor epilepsy* or *temporal lobe epilepsy,* and are characterized by 1) an aura--which has a hallucinatory or illusionary component to it, and 2) an alteration of behavior which terminates with amnesia for the event. The subjective component may consist of illusions (distortions of perception, hallucinations (visual, auditory, gustatory,olfactory, or vertiginous), dyscognitive states (deja vu-inappropriate feeling of familiarity, jamais vu-inappropriate feeling of strangeness), or affective experiences (usually anxiety). Rage or intense anger is unusual in complex partial seizures. The objective components are usually referred to as automatisms, and may consist of lip smacking, chewing movements, hand wringing, foot shuffling or complex unwitting activities like walking in a daze, undressing in public,etc. These activities are usually automaton-like, and the person will attend to questions, but respond not at all or in a confused fashion. Violence is unusual, but if the person is physically directed, will usually resist. The postictal state is usually one of confusion.

The subjective components of complex partial epilepsy may occur without the objective components, and in that case is called a simple partial seizure. All forms of partial seizures may generalize. The significance and importance of classifying seizure type is that many forms of partial seizures (even if they progress to generalized seizures) are amenable to treatment surgically, whereas generalized seizures are not. Also, different seizure types may respond to different medicines.

Generalized Seizures are what most people think of when they think of epilepsy. They are all accompanied by loss of consciousness, so the terms 'simple' and 'complex' do not apply. The entire cerebral cortex is involved, and, if the seizure is motor, all muscle groups are involved. The common types are: grand mal, petit mal, Lennox-Gastaut, juvenile myoclonic, infantile spasm, and atonic.

Grand Mal Epilepsy may be tonic (increase muscle tone, spasm), clonic (contractions), or tonic-clonic (contractions alternating with spasms). This is the most prevalent form of generalized epilepsy, and is a 'classic' seizure, in the public mind.

Petit Mal Epilepsy is often called Absence seizures, and consists of a brief loss of consciousness with or without brief motor movements. These people seem to be daydreaming, and if there is no motor component, may respond to questions with a blank stare. The motor movement is usually very subtle, and may consist of a slight twitch at the corner of the mouth. The seizures are usually very brief, lasting seconds rather than minutes, reinforcing the opinion of the observer that they have been caught not paying attention.

<u>Lennox-Gastaut syndrome</u> is a mixture of several seizure types, beginning in early infancy and usually accompanied by retardation.

<u>Atonic Seizures</u> are also called drop, akinetic, or astatic seizures. They usually consist of a momentary loss of muscle tone, the person often regaining consciousness before falling, and giving the appearance of a stumble.

<u>Juvenile Myoclonic Epilepsy</u> and <u>Infantile Spasms</u> are very rare types of generalized seizures and require specialist interpretation, often involving other motor involvements.

<u>Altered Metabolic States</u>, such as extreme hypoglycemia or hyperthyroidism, can result in a generalized seizure. These events are usually not repetitive in the sense that epilepsy is, but occur at the time of the acute metabolic disorder.

DIAGNOSIS of seizure disorders starts with the history, including a detailed description of the seizure itself. This is one of the reasons the above discussion is important. The history often is obtained from teachers or other interested bystanders.

The next step in certifying the diagnosis is an electroencephalogram (EEG). This is a measurement of electrical potentials on the surface of the brain. The leads (amplifiers) are placed on the scalp in a predetermined pattern. These are from 8-16 in number, and the voltage between each pair of leads is compared to that between other pairs. The EEG patterns help determine the type of seizure.

TREATMENT of seizure disorders is accomplished with one of the following anticonvulsant drugs (also referred to as anti-epileptic drugs or A.E.D.):

Barbiturates (Phenobarbital) -- one of the earliest and most effective anticonvulsants. However, side effect of sedation has caused it to be replaced by one of the subsequent drugs. Still often used for rapid effect in status epilepticus.

Phenytoin (Dilantin) -- discovered early and soon replaced or became adjunct to Barbiturates. Has more side effects, but still superior for control of some types of Partial Seizures. Has been virtually replaced by subsequent, safer drugs.

Benzodiazepines (Valium) -- Very effective fast acting anticonvulsant, quickly became substitute for phenobarbital in status epilepticus. Also strong anti- anxiety component led to wide use as sedative. Due to propensity for addiction, widely replaced by SSRIs.

Carboxamide (Tegretol) -- Effective anticonvulsant with good safety profile. Side effects usually controllable with monitoring. Also effective mood stabilizer and controller of neuropathic pain.

Valproate (Valproic Acid) -- Similar effective range and safety profile as carboxamides. Side effects slightly different and require monitoring.

There has been an explosion of variations on these basic chemicals and a few others. People who have a need to know are advised to find one of the many excellent monographs on anticonvulsant or psychotropic pharmacology; as many of these drugs effective on the neurological system have also found many uses as psychopharmacological agents.

VASCULAR DISORDERS

Stroke is an outmoded term for vascular occlusion to part of the brain. The blockage of the vessel is of like pathology in heart attacks and in strokes; damage to the blocked tissue is similar, causing an area of cell death beyond the obstruction. The phenomenon can also occur due to bleeding as well as blockage. The result of the damage is more likely to be fatal in heart attacks, leaving a proportionally larger group of survivors with chronic brain damage for whom care is necessary. As a result the term *stroke* persists in folk consciousness. Consequently, a relatively large number of brain damaged stroke survivors required chronic care, as a consequence of which a large data base of clinical information accumulated. This could then be correlated with autopsy findings which had been preserved for many years as scarring in the healed infarcts.

This comparison was incorporated into a body of knowledge characterized as *functions of brain centers*. As diagnostic techniques burgeoned, from EEGs, spinal taps and X-ray to pneumoencephalograms and ultrasounds, then CAT, MRI and MRA, the location and nature of lesions could be determined with greater precision; and on live patients. Also, areas involved in mental activity can be studied with PET and fMRI scanners. In the future, it is expected that neural connections creating feedback loops, monitoring and modulating systems, as well as webs can surely be analyzed.

The most obvious effects of a stroke consist of memory loss and paralysis on the other side of the body.

Right brain strokes	Left side strokes
vision problems	speech / language problems
impulsive, inquisitive behavioral style	slow, cautious behavioral style
L side neglect	
loss short term memory,	
persistent belief in abilities no longer present.	

Aneurysm - a widening or ballooning of an artery, due to weakness in the wall. When it occurs in the cerebral circulation damage is caused by pressure compression since the skull is a closed space, or by rupturing and resulting in a hemorrhagic stroke or depriving other tissues of nutrients and oxygen. and do damage either by pressing on surrounding structures (parts of the brain), Such structures are usually due to congenital weakness in the wall of the artery, or can be acquired by atherosclerosis or other disease which can destroy the vessel wall.

SYNDROMES [Merriam-Webster]

a group of symptoms or signs typical of a disease, disturbance, condition, or lesion in animals or plants. In the history of medicine, the cluster of signs and symptoms and, sometimes, phenomena which are related to a disease state may have been, and sometimes still is, the only indication of that disease. As the causes are discovered, more and more syndromes will have a recognizable disease as the cause. Until then astute clinicians will suspect a disease state as one or a few syndromic components are recognized.

The words *disease* and *syndrome* are used quite flexibly today. Language is a growing changing thing and it is always wise to have a clear understanding of what is meant when words are used; in this case we will use <u>syndrome</u> to mean a cluster of signs (physical characteristics observed by the clinician) or symptoms (feelings reported by patient); <u>disease</u> will be used to indicate the cause of the signs and symptoms.

This distinction becomes especially important in conditions (like minor physical anomalies -- sometimes referred to as dysmorphism) which are fairly frequent and can either be inconsequential or may be associated with abnormal neurological development (especially if found in typical combinations - syndromes).

It has been estimated [Marden] that newborn babies with one minor anomaly did not have an appreciable increase in major anomalies; babies with two minor defects showed an increase in major defects to five times the general group; and 90% of those born with three or more minor anomalies had a major anomaly, as well. Thus, several minor anomalies in the same individual is unusual and often indicates the likelihood of a serious problem in prenatal development.

The minor anomalies of which we speak include complex areas like the face, ears, hands and feet. Whorl patterns in the hair, spacing of eyes, elevation or depression of lateral canthi, malformed (not unfolded) ears, placement of nasal bridge, shape of mouth, webbed digits, crease patterns of palm and sole, as well as fingerprint patterns are characteristic minor anomalies. A good atlas [Smith] is worth studying for familiarization with these patterns. A book like Smith's also contains an index indicating the combinations of minor anomalies constituting syndromes.

CHROMOSOMAL ABNORMALITIES

In 1866, John Down wrote *Observations on the Ethnic Classification of Idiots,* in which he described the findings which typified the syndrome. One of the characteristics he noted was the similarity to the Mongoloid race, resulting the syndrome being called Mongolian Idiocy, then Down Syndrome.

In 1950, Karyotype techniques (chromosomal analysis) was developed and Jerome Lejeune discovered that the cause of Down Syndrome was an additional chromosome 21, leading to the naming the disease Trisomy 21. Since then many additional Chromosomal syndromes have been named, the ranging from metabolic abnormalities to trisomies with the complexity of Down's. Trisomy 8, 9, 13, 16, 18 and 22 are less common than 21, but appear with regularity in institutions that care for children with problems.

NEUROCUTANEOUS DISORDERS

The relationship between the nervous system and the skin is more than the fact that nerve endings in the skin form receptors for the sense of touch. The connection is subtle and intimate, beginning during embryonic stages. In the early embryo, the tissue which becomes the nervous system originates when the outer dorsal layer (that tissue which becomes skin on what becomes the back of the organism) invaginates into the substance of the embryo and gets pinched off, forming a neural tube, which develops into the brain and spinal cord.

Genetic, dominant -- *Neurofibromatosis Type I* -- >6 cafe au lait spots larger than 5 (prepubertal) -15mm (postpubertal). Also, rubbery subcutaneous nodules. Risk for developing numerous benign or malignant tumors.

Neurofibromatosis Type II -- associated with acoustic neuroma and other tumors.

Tuberous Sclerosis -- adenoma sebaceum, epilepsy, and retardation. Fibromas can develop in other organs.

Nevoid Basal Cell Carcinoma, Variegate Porphyria, LEOPARD Syndrome, and Osler-Weber-Rendu disease are examples of other genetic dominant conditions.

Genetic, recessive -- *Ataxia-telangiectasia* -- is a condition of progressive ataxia (poor coordination) and telangiectasia (small dilated blood vessels)

Sturge-Weber -- a port wine stain in the cranio-facial area. Usually associated with seizures and cerebral calcifications. The port wine stain is deep red and irregularly shaped. It is not to be confused with strawberry hemangioma, a bright red lesion which looks like a strawberry and usually resolves by itself.

INFECTIONS

Shingles -- A vesicular eruption along the distribution of a nerve root. The vesicles are tiny fluid filled blisters, each on a red base identical in appearance, but not distribution, to chicken pox. The causative virus is known as Varicella Zoster Virus, Human Herpes Virus Type 3 and causes chicken pox in previously uninfected people (usually children); following recovery the virus becomes dormant in a nerve root and re-erupts, usually in elder years as shingles.

Meningitis - inflammation of the nervous system lining (meninges). Usually due to infection with pyogenic (pus producing) bacteria, though it can be caused by other microorganisms.

Encephalitis - inflammation of the brain usually from infection most often by a virus, but occasionally other agents.

The early symptoms are fever, vomiting, headache and irritability. Though many of these symptoms are also induced by many septic agents in systemic locations, the frequency with which it accompanies inflammation of or pressure on the brain is the reason for including them in the clinical examples, which illustrate the functional connections of neurophysiology. Diagnosis requires many more dimensions of symptom frequency and association than simple lists like this illustration **allows**.

Encephalopathy (L. encephalo *brain* and pathy *disease*) refers to any of various diseases of the brain, including the above. Some forms of encephalopathy can be caused by metabolic disorders, toxins (internal or external), hypoxia, hypertension, trauma, etc.

Developmental Disorder [REYNOLDS]

an impairment of the growth and development of the brain or central nervous system. In the U.S. school age population, the three most frequent Developmental Disorders [CENTERS FOR DISEASE CONTROL], in decreasing order of frequency are: Mental Retardation, Cerebral Palsy, and Autism Spectrum Disorders. Other conditions believed to be the outcome of some abnormal process that unfolded as the brain was developing *in utero* or in the young child, include attention deficit/hyperactivity disorder (ADHD), epilepsy and Tourette syndrome.

Mental Retardation [American Psychiatric Association] -- a generalized disorder which has appeared in childhood characterized by impaired cognitive functioning, and involving deficits in at least two adaptive behaviors. The word is still used in some government departments, medical terminology and other places, but is being replaced by labels considered less offensive in many advocacy circles as well as the DSM-5 [American Psychiatric Association] which has replaced "Mental Retardation" with a new name:

"intellectual disability (intellectual developmental disorder)". The delays may be associated with other signs and symptoms as described in Chromosomal Syndromes and be referred to as syndromic mental retardation; or may occur without association and be called non-syndromic mental retardation.

Cerebral Palsy -- discussed in section on abnormal motor movements.

Autism Spectrum Disorders [Johnson] -- characterized by impaired social interaction and communication. Also has repetitive, stereotypic behavior, and that the symptoms are apparent before the age of three. The spectrum of autism ranges in functioning from higher (Asperger's Syndrome) to lower (Pervasive Developmental Disorder n.o.s.)

MYELODYSPLASIA

The abnormal development of the vertebral column or spinal cord or both. There are three degrees of severity: *spina bifida occulta, meningocele, and myelomeningocele.* The normal anatomy of the spine consists of the spinal cord, which is the composite of nerve fibers descending down the back from the brain, the vertebral column, which is the series of bones that encase the spinal cord, the muscles lying along the vertebral column, which help hold us upright, and the overlying skin. Central to understanding myelodysplasia is the understanding of the relationships between the spinal cord and vertebral column. The stack of vertebrae, known as the vertebral column (bony spine), consist of a strong mass of bone, known as the vertebral body, and posterior (towards the back) projections of bone, known as the laminae, which form an arch, through which the spinal cord travels. The spinal cord is covered with a protective membrane called the meninges (pia mater, arachnoid mater and dura mater).

The spine consists of 24 articulating (connected by joints) vertebrae and 9 fused vertebrae. The articulations allow some movement (bending, twisting), and provide spaces for egress of the peripheral nerves from the spinal column.

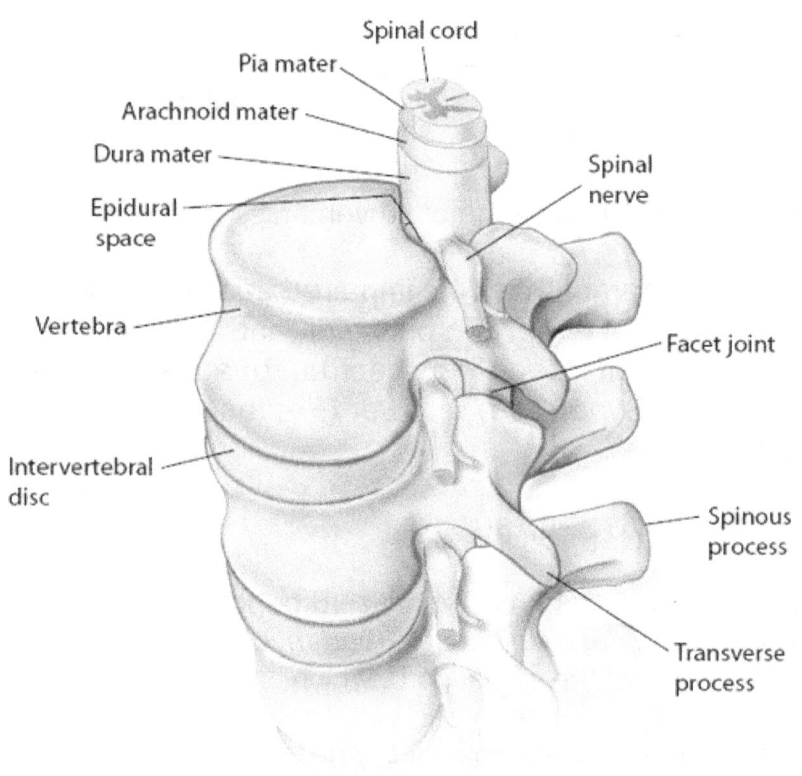

Segment of spinal column and spinal cord
Courtesy Columbia University Department of Neurological Surgery

spina bifida occulta is the mildest form of myelodysplasia, involving only failure of fusion of the posterior portion of the vertebral column, which results in an incomplete arch. There is no neurological deficit, nor external signs of the disorder (thus the name *occulta*, L. for hidden). It occurs in about 5% of newborns, and usually occurs in the lumbosacral region. Since the arch is incompletely formed, fatigue fractures occur at the location where the arches join each other (pars interarticularis), and these fractures are known as spondylolysis (spawn-dee-low-lie-sis). There is also an increased likelihood of the vertebrae slipping over each other, known as spondylolisthesis (spawn-dee-low-lis-thee-sis).

meningocele is a more severe form of myelodysplasia, involving failure of fusion of the posterior arch of the vertebrae, and herniation of the meninges (membraneous covering of the central nervous system). Usually the meninges are intact (not ruptured) and the skin is usually covering the defect. Sometimes a bulge is apparent, leading to the possibility of trauma. There are other forms of meningoceles, but this is the most common.

myelomeningocele is the most severe form of myelodysplasia, and consists of the boney defect (failure of posterior fusion of the vertebral arch), herniation of the meninges (sometimes ruptured, or open), frequently with the skin open over the defect, and abnormal neural elements. The degree of abnormality to the neural elements is related to the level of the lesion.

As a guide, most orthopedists estimate that if the
myelomeningocele level is at:	the resultant handicap will involve:
sacrum	community ambulation
L 4-5	household ambulation
L 3	nonfunctional ambulation
thoracic	non-walker

As we can see, the higher on the spinal cord the lesion is, the more neural tissue is involved. Other problems that arise in myelomeningocele include renal problems, scoliosis, hip dysplasia, and occasionally the child will have:

Arnold-Chiari malformation which includes hydrocephalus, profound neurological deficit with motor loss and deficient or absent sensation, and bowel and bladder paralysis.

The myelodysplasia syndromes are related to folic acid deficiency in the mother's diet early in pregnancy. The cause of the Arnold-Chiari malformation is not known.

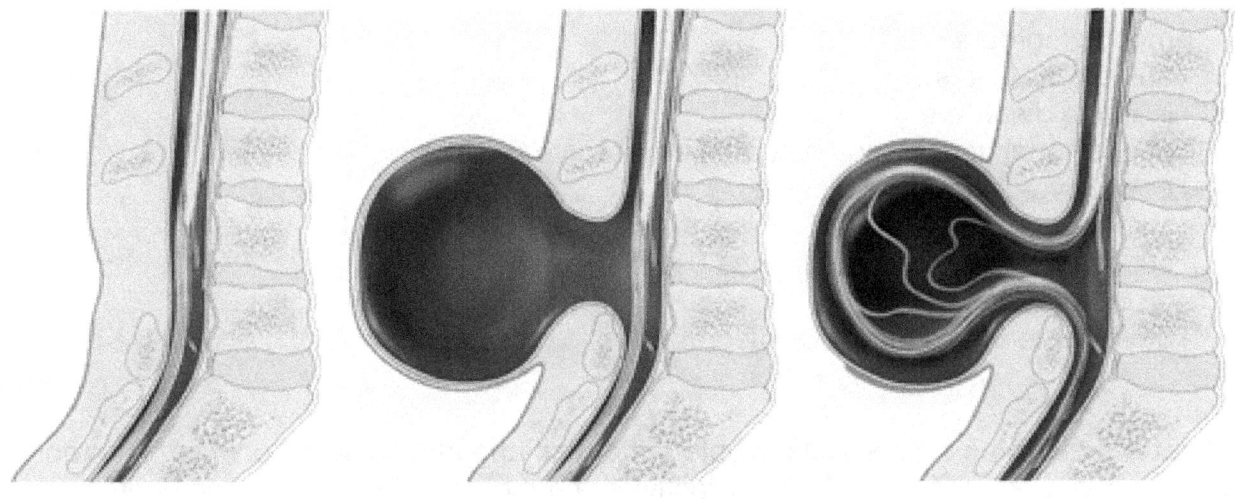

Spina bifida occulta   Meningocele   Myelomeningocele

Courtesy of Centers for Disease Control and Prevention

OTHER ABNORMALITIES ASSOCIATED WITH PRENATAL INFLUENCES

Attention Deficit/Hyperactivity Disorder (ADHD)
The classification of subtypes in this disorder is undergoing revision. Generally all subtypes are characterized by:
- inattention
- easy distractibility
- disorganization
- procrastination
- forgetfulness.

The *ADHD* subtype is strongly characterized by hyperactivity and impulsiveness leading to more frequently associated conduct disorders.
The *ADD* subtype is associated with more anxiety.

Comorbidity (conditions existing simultaneously with or independently of index condition[3]):
- mood disorders
- anxiety disorders
- learning disorders
- oppositional defiant, conduct, and aggressive disorders
- obsessive-compulsive disorder
- sleep disorders
- substance abuse and substance use disorders
- tic disorders
- developmental coordination disorders

Tourette syndrome consists of multiple motor tics and at least one vocal tic. The motor tics should not be confused with hyperactivity and the vocal tic need not be coprophilic. Tourette's syndrome can be comorbid with ADD and/or OCD.

Motor tics are sudden, rapid, nonrhythmic movements.

hydrocephalus - abnormal accumulation of cerebrospinal fluid (CSF) in the ventricles, or cavities, of the brain. The result of the fluid accumulation is increased pressure, pressing the brain tissue against the skull resulting in damage, seizures, and eventually death.

---

[3] the reason diagnosis should be conducted by clinicians familiar with these disorders.

Part III
SELECTED FUNCTIONS

Afterword

A number of topics do not fit easily into the format of this book:
- consciousness
- memory
- personality

COLLECTIVE CONSIDERATIONS

The topics all represent functions of the Mind. There are others, like Intelligence, Emotion, Volition, but these will have to do to represent those functions not connected primarily to physical neurological functions. This book has primarily been about the brain, a physical structure; the mind seems more in the category of psychology, but there are reasons to attend to the connections these three phenomena have with the brain. Also Neuroscience has been productive in increasing knowledge in these abstract areas; making it incumbent to identify connecting bridges from concrete to abstract, particularly as our tour of anatomic divisions, in Part I has brought us to these limbic system connections - a good linking point.

A reason they don't seem to fit is that they resist definition. On the other hand, the reason they must be included is because they are clearly connected to the brain. We shall address each of these topics individually and try to find a way into the Mind - Matter difficulty.

DEFINITION, LEXICAL[4], DICTIONARY[5] - The main reason they don't seem to belong is they represent abstract, rather than concrete, things. Concrete beings are visible, tangible and otherwise knowable by the senses. They can have their parts observed, enumerated and described. Abstract concepts are often vague, ineffable and sometimes called *Universals*, because, as concepts they provide properties to a list of concrete particulars of which they are an instance. Essentially, they can have their parts described only as they are perceived - usually as a model, synonym list, a series of

---

[4] This word is used as the common meaning of dictionary: the book of definitions, spellings. parts of speech, synonyms and etymologies.

[5] This disambiguation of the word, Dictionary, is meant to specify the section which catalogs the words alphabetically, for definition; not the general meaning of the Book and all it's lexical contents

processes or examples (usually of effects[6]), or by analogy (which is not much different than synonyms).[7]

Even the best lexical definitions tend to be circular. Definitions are usually descriptive and/or enumerate or pen in the characteristics which form the substance of the term; but most dictionaries are designed to demonstrate current usage of said term. According to some modern analytic philosophers it is current usage that gives the term meaning [WITTGENSTEIN]. It is not possible to define them ostensively by pointing, since they are not physical structures.

The alternative seems to be describing their functions, but there problems seem to not only be plentiful and even multiply. They do not have clear linear connections from one synapse to the next; not even from a receptor to a nucleus, not clearly ending in an area known to specialize in that topic[8].

The difficulty of their definition has been characterized: [SAINT AUGUSTINE][9]. It is not that we can't agree on their meaning, but that any meaning that is commonly used, for ordinary purposes is incomplete. A selected list of definitions types:
- lexical - (dictionary, thesaurus, etymology)descriptive of common use, not prescriptive
  theoretical - assumes both knowledge and acceptance of the theories upon which a specific discipline depends.
  creates a hypothetical construct[10].
- persuasive - purports to be lexical; with stipulative factors
  e.g., (an Intellectual is someone who forgot how to use tools.)
- stipulative - a term is given a specific meaning[11].
- operational - "doing" or "how-to" ex., recipe, music score
- ostensive - miming or pointing [Johnson, William Ernest]

DEFINITION, LEXICAL, SYNONYMS -
These words are usually offered as alternative uses which have a meaning similar to the index word. A thesaurus is more extensive, offering the range of the senses of meaning a word ostensibly has. It offers a platform of meanings which should support

---

[6] Many proofs of the existence of God enumerate the Wonders of Creation (hypothetical effects of a hypothetical cause)

[7] In Aristotle's view, universals exist only where they are instantiated; they exist only in things, never apart from things.

[8] such as retina-lateral geniculate-occipital cortex.

[9] "What then is time? If no one asks me, I know what it is. If I wish to explain it to him who asks, I do not know." Saint Augustine, *Confessions*, Book 11

[10] intelligence def. requires comprehensive def.of mind & reality; idealisms like ideal gravitational point, center of gravity

[11] for the purposes of argument or discussion in a given context.

the broadest definition. A way of dealing (heuristics) with a list of synonyms has been addressed previously [STEVKO, 2013].

DEFINITION, LEXICAL, ETYMOLOGY - the origin of words and the way in which their meanings have changed throughout history[HARPER]. A way to explore meanings intended by purported originators of the word.

Another method that is a little less authoritarian than use of dictionaries and thesauri is coming into vogue and can be called Mental Archaeology. This technique seems close to digging around the minds of people who ostensibly coined the term in question. That excavating requires a lot of imagination to have many options to consider, and a lot of restraint to not put our preconceived notions into the mouths of people who can't talk back. We also have to make ourselves available to scholars in the field who have built up a taste for what was authentic to the language and the time.

BIOLOGICAL CONNECTION - Despite difficulties of definition these functions of the mind do belong in this work. Careful historical records of the malfunction correlates functional loss with identification of damaged parts, especially to losses of: learning (hippocampus, cerebellum) memory (amygdala), personality (frontal lobe). They also play parenthetic, but important roles in ADD, a plethora of agnosia disorders, temporal lobe epilepsy, and any number of disorders which have been written off as psychiatric when neurological connections are not recognized. Observed increase in functions when a proposed site is stimulated during surgery. The other connections noted between mental functioning and the brain is increased activity on PET or fMRI scans.

It is not apparent which of the topics we recognize are related, nor what the definitive functions of each phenomenon are. Biological roles are more easily studied in animals, but their interpretations are widely subject to anthropomorphizing, which creates another circular situation. However, [CHARLES DARWIN] managed to explore animal behavior and initiate many behavioral evolutionary disciplines, successfully.

Rather than understanding how these phenomena arrived in the mind, or how they're related or from where they seem to have arisen, they are often preferentially reified with extra-empirical, sometimes supernatural qualities. Be that as it may, it seems important to point out that these states depend on connections no matter how their ultimate import is interpreted; especially as modern neurophysiology, neuropsychology, and neurophilosophy does seem to fill in the gaps and offer some rather primitive, at this time, pathways to connect these states to our experience in the world.

When the phenomena of consciousness, memory and personality were recognized and named, the disciplines of neurophysiology, neuropsychology, and neurophilosophy did not yet exist. That is not to say that the neural connections contain all the information we seek, but it is demonstrably connected and the biological part is a part we cannot ignore. The failures of mental function due to damage of brain tissue has been apparent, but the loss is not linearly related, but integral through association areas that are comprised of an apparent chaos of neuronal networks awaiting disentanglement.

The use of terms like phenomena in a section which discusses binary oppositions like *abstract / concrete*, *real / ideal*, or even functional oppositions like *mind / brain* is going to run into separations like phenomenon / noumenon and depending on inclination may even confront philosophical oppositions to oppositions.[SCHOPENHAUER] Without involving a different dimension, an abbreviated glossary is included in the Back-matter preceding the bibliography.

And there we have it. The entire complex of information which does not fit our knowledge is certainly due to insufficient information. There are some other groups, Dualists (mind vs. matter), who from Plato through Descartes and still to quantum physics where the wave / particle dualism of matter remains unresolved, that deny mind-body connections can be made; there are other disciplines, as well, that claim clear connections to the mind - hunting and gathering the unknown pieces. A dualistic attitude would not allow a new synthesis of knowledge; indeed, would make it more difficult to attain.

As in all of science, when we do not know the answers, we begin speculating within the interstices of what we do know and attempt formulating a narrative that makes sense. This happened when the alchemists were faced with beakers that held substances known by bewitching names, like quicksilver (mercury) and aqua fortis (nitric acid). It took many chemical reactions collected as data, codified as information and annealed with imaginative thinking to progress from observed effects to categorize a power (positive or negative) to interact as a quality called *valence*; then eventually visualize those powers as electrons, protons and neutrons.

The same thing happened when the quantum physicists charmingly named a hypothetical particle, *the quark* (a name derived from James Joyce's *Finnegans Wake);* and further named particular types of quarks as *charm* and *strange.* It was not only in science that this type of thinking took place. All peoples everywhere try to formulate what is difficult into comprehensible disciplines which are then canonized into bodies of knowledge. The scientists took credit for the empirical aspect of this paradigm and called it scientific method; the philosophers applied it to metaphysical knowledge and called it philosophy; theologians applied it to supernatural knowledge and called it religion.

Indeed, the tendency to story tell as an explanatory device, from toddlers, through childlike animists to imaginative narrators, and ending with mature speculators may well be inbred as language is thought to be. [CHOMSKY]

The scientists have, of necessity, modified their canon when new information is better at explaining the unknowable. Further, certain claims made by post-modern philosophers (Quine, Kuhn, Derrida) that all knowledge is provisional, composed of incomplete concepts subject to revision as new, reliable knowledge is uncovered. Brutus explained, "GOOD REASONS MUST OF FORCE GIVE PLACE TO BETTER".[SHAKESPEARE]

INDIVIDUAL CONSIDERATIONS

CONSCIOUSNESS

definition - lexical, dictionary
consciousness - The state of being awake and aware of one's surroundings [OED, 2013]

Conscious - refer to an individual sense of recognition of something within or without oneself. Implies to be awake or awakened to an inner realization of a fact, a truth, a condition, etc. [RANDOM HOUSE DICTIONARY]

conscious - (adj.) perceiving, apprehending, or noticing with a degree of controlled thought or observation...[MERRIAM-WEBSTER]

definition - lexical, synonyms -

Conscious	Aware, Cognizant
Aware -	lays the emphasis on sense perceptions insofar as they are the object of conscious recognition
Cognizant -	lays the emphasis on an outer recognition more on the level of reason and knowledge than on the sensory level alone
Know	Comprehend, Understand
To know	is to be aware of something as a fact or truth
comprehend	to know something thoroughly and to perceive its relationships to certain other ideas, facts, etc.
To understand	to be fully aware not only of the meaning of something but also of its implications
percipient	become aware through the senses.

covers senses of meaning from knowledge by sensation through cognition and adds ranges of degree of understanding.

definition - etymology -
Our *mot du jour*, derived from <u>conscious</u> arose in this form from Latin *conscius* (*con-* "together"[12] + <u>*scio*</u> "to know")
There were, however, many occurrences in Latin writings of the phrase *conscius sibi* in which *Sibi* means "to oneself"

---

[12] Although its sense in compounds of Latin derivation is often obscured, its meanings usually are: together, with, etc (*combine, compile*) ; similar (*conform*); extremely, completely (*consecrate*)

A similar word from the time is: Gk. syneidos. : .
syn-, of like meaning or like name. — Gk
Eidos, a Greek term meaning "form," "essence," "type," or "species". See Plato's theory of Forms and Aristotle's theory of universals (described as types [particular], properties [attribute, quality], relations).

etym.: Latin *conscius* (*con-* "together" + *scio* "to know"), but the Latin word did not have the same meaning as our word—it meant *knowing with*, in other words *having joint or common knowledge with another*.[11] There were, however, many occurrences in Latin writings of the phrase *conscius sibi*, which translates literally as "knowing with oneself", or in other words *sharing knowledge with oneself about something*. This phrase had the figurative meaning of *knowing that one knows.* It can also be construed to agree with the sense, *speculation*, the kind of thinking common in philosophy and hypothetical exploration. [STEVKO, 2014].

BIOLOGICAL CONNECTION -
biological factors [ZEMAN, 2001]
topical
    renewed respectability among psychologists
    rapid progress in the neuroscience of
        perception
        memory
        action
    advances in artificial intelligence
    dissatisfaction - dualistic separation mind and body
The understanding of states of consciousness has been transformed by
    the delineation of their electrical correlates
    of structures in brainstem and diencephalon which regulate the sleep-wake cycle
    of these structures' cellular physiology and regional pharmacology.
Clinical studies have defined pathologies of wakefulness:
    coma
    the persistent vegetative state
    the 'locked-in' syndrome
    akinetic mutism
    brain death
Interest in the neural basis of perceptual awareness has focused on vision.
    Increasingly detailed neuronal correlates of
        real and illusory visual experience are being defined
    Experiments exploiting circumstances in which
        visual experience changes while
        external stimulation is held constant
            are tightening the experimental link
            between consciousness and its neural correlates.

Aesthetic experiences are experiments [LEHRER] in that a physical stimulus
applied creatively to a sentient being results in a visceral reaction, as did:

- Paul Cezanne, vision
- Igor Stravinsky, sound
- Auguste Escoffier. taste
- Marcel Proust, smell
  before the development of neuroscience, and as did:
- Walt Whitman
- George Eliot
- Gertrude Stein
- Virginia Woolf, in their direct appeal to consciousness.

Multiple other artists have continued the effort, and some neuroscientists have
studied the relationship directly [ZEKI] [RAMACHANDRAN].
Work on unconscious neural processes provides a complementary approach.
    'Unperceived' stimuli have
        detectable effects on neural events and
        subsequent action in a range of circumstances:
            blindsight provides the classical example
    Other areas of cognitive neuroscience also
        promise experimental insights into consciousness
            in particular the distinctions between
                implicit and explicit memory and
                deliberate and automatic action.
Overarching scientific theories of consciousness include
    neurobiological accounts which specify mechanisms[13] for awareness,
    theories focusing on the role played by conscious processes
        in information processing, and
        theories envisaging the functions of consciousness in a social context.
Whether scientific observation and theory will yield a complete account of
consciousness remains a live issue.
Physicalism
functionalism
property dualism
dual aspect theories
attempt to do justice to three central, but controversial, intuitions about experience:
    that it is a robust phenomenon which calls for explanation,
    that it is intimately related to the activity of the brain and
    that it has an important influence on behavior.

---

[13] anatomical or physiological

MEMORY

definition - lexical, dictionary
- The faculty by which the mind stores and remembers information
- Something remembered from the past (a recollection)
- The part of a computer in which stores data or program instructions for retrieval.

definition - lexical, synonyms

definition - lexical, etymology
Middle English: from Old French *memorie*, from Latin *memoria*, from *memor* 'mindful, remembering'.
noun of quality from *memor* "mindful, remembering," from
PIE root **(s)mer-* "to remember"
(Sanskrit *smarati* "remembers,"
Avestan *mimara* "mindful;"
Greek *merimna* "care, thought," *mermeros* "causing anxiety, mischievous, baneful;"
Serbo-Croatian *mariti* "to care for;"
Welsh *marth* "sadness, anxiety;"
Old Norse *Mimir*, name of the giant who guards the Well of Wisdom;
Old English *gemimor* "known," *murnan* "mourn, remember sorrowfully;"
Dutch *mijmeren* "to ponder").

Wotan's ravens: Huginn (Thought) and Muninn (Memory) flew throughout the world and reported its activities to the chief god; the Scandinavian legend thereby initiating the idea of information from the external world being processed by thought and memory.

BIOLOGICAL CONNECTION -
Brain structures involved in memory are in two wide groups: cortical and subcortical. Though those two groups seem to comprise most of the brain, this categorization has the merit of subdividing different types or methods of handling memory.

Cortical -
Frontal Lobes - coordination of information, working memory, prospective memory
Temporal Lobes - autobiographical, familiarity, language, long term memory
Parietal Lobes - short term memory, spatial memory
Occipital Lobes - visual memory

Subcortical -
Hippocampus - cognitive maps, encoding
Cerebellum - procedural (motor&cognitive) memory
Amygdala - memory of fear conditioning, memory consolidation
Basal Ganglia - motor memory

memory physiology

neuroanatomy of memory		
hippocampus	spatial learning declarative learning explicit memory memory consolidation	input and output to entire cortex incl. 2° & 3° sensory areas that have already processed the information
amygdala	emotional memory	
striatum	attention short term memory	
mammillary bodies	episodic memory	

Damage to an area is actually responsible for the observed deficit
    could implicate a specific area
    could involve adjacent areas
    could involve a pathway traveling through.
Learning and memory are
    not solely dependent on specific brain regions
    attributed to changes in neuronal synapses. Learning results from changes in the strength of the synapse [Cajal].
    thought to be mediated by long-term potentiation and long-term depression.
memory enhancement factors
    help the storage of recent experiences
        direct injections of cortisol or epinephrine
        excitement stimulates hormones that affect the amygdala
    hurt memory storage
        Excessive or prolonged stress (with prolonged cortisol)
        damage to amygdala
            emotionally vs. non-emotionally charged words
                no preference for more likely to remember

PERSONALITY

definition - lexical, dictionary - The combination of characteristics or qualities that form an individual's distinctive character [OED]

definition - lexical, synonyms
characteristic - an identifying feature or quality
character - The mental and moral qualities distinctive to an individual:
quality 1 - The standard of something as measured against other things of a similar kind; the degree of excellence of something:
    2 - A distinctive attribute or characteristic possessed by someone or something

definition - lexical, etymology - Middle English (in the senses 'character, disposition' and 'particular property or feature'): from Old French *qualite*, from Latin *qualitas* (translating Greek *poiotēs*), from *qualis* 'of what kind, of such a kind'.

history - The word "personality" originates from the Latin *persona*, which means mask. In the theatre of the ancient Latin-speaking world, the mask was not used as a plot device to *disguise* the identity of a character, but instead was a convention employed to represent or *typify* that character.

older references to the topic of personality usually used words like person, character. word not common until Renaissance when Enlightenment humanized sense of self from network of social roles to an individualistic concept.

BIOLOGICALLY BASED - events

Phineas Gage - frontal lobe injury resulted in personality change [Damasio, H] [Damasio A.R.]
Eadweard Muybridge—an early case of frontal lobe injury leading to mental changes
Samantha Fox - another case of frontal lobe injury leading to mental changes

Anatoli Bugorski—particle-accelerator proton beam passed through head.
   reported flash of light and subsequent seizures.
   did not report any behavioral changes.
   probably through temporal, not frontal, lobe
astronauts more vulnerable to cosmic rays when in orbit also reported seeing flashes of light.

Primary Polydipsia (compulsive water drinking) - occurs with with head injury or other conditions that damage the anterior pituitary gland. Children who had benign conditions that affected the gland developed polydipsia, which persisted despite correction of the causative conditions, and stabilization of fluid balance homeostasis. The question of physiologic adaptation, resembling habit, remains unresolved. [STEVKO 1968]

BIOLOGICALLY BASED - personality theories
Hans Eysenck
    a causal theory of personality based on activation of reticular formation and limbic system.
Jeffrey Alan Gray [CORR] - three brain systems that all differently respond to rewarding and punishing stimuli.

System	mediates	personality traits
Fight-Flight-Freeze	the emotion of fear (not anxiety) and active avoidance of dangerous situations	fear-proneness and avoidance
Behavioral Inhibition	the emotion of anxiety and cautious risk-assessment behavior when entering dangerous situations due to conflicting goals	worry-proneness and anxiety
Behavioral Approach	the emotion of 'anticipatory pleasure,' resulting from reactions to desirable stimuli	optimism, reward-orientation, and impulsivity

Cloningen Claude Robert
    associates personality models with chemical components
        novelty seeking - dopamine
        harm avoidance - serotonin
        reward dependence - norepinephrine
    studied by EEG, PET, fMRI

genetics - It is suggested that heredity and environment interact to determine one's personality.
HUMAN GENOME PROJECT - Mutations were identified that suggested the existence of genes determining several personality traits. [FITZGERALD DA]
TWIN STUDIES - Researchers from Edinburgh University found that identical twins were twice as likely as non-identical twins to share the same personality traits, suggesting that their DNA was having the greatest impact. [ARCHONTAKI]

We tend to speak of these phenomena, not only here but in conversation and consultation as if they were concrete objects. As new information is found, It will be important, to not only change our concepts, not as old friends who behave differently, but as barely known acquaintances for whom we had affection and are now willing to relate on a different level. After all, phenomena are known authentically only when we are willing to allow them to show all their elements freely, rather than fitting them into a preconceived notion

HISTORY OF UNDERSTANDING BRAIN FUNCTION

General: The functions of the various parts of the body have been learned laboriously over a long time. Two methods predominated in gathering this information:
- inductive reasoning - particular to universal - when a lesion occurred in a body part, the lost function was considered the function for which that part was responsible.
- deductive reasoning - universal to particular - when people reflected on body functions, the considered function was added to the list that had already been imagined or compiled.

loss - The functions of the brain were learned, as were most other body organ functions, by cause and effect associations. When a part of the nervous system is damaged, the role that is resultantly lost is attributed as the function of that part.

TIMELINE
- c. 1500 BC - The *Edwin Smith Surgical Papyrus* - A surgical papyrus was a compilation of 48 cases, including 27 head injuries. In 1862, it was purchased in Luxor by Edwin Smith (1822 – 1906), after whom it was named.
- 460-379 B.C *Hippocrates* - Greek restriction on dissecting
    - discusses epilepsy as a disturbance of the brain
    - states that the brain is involved with sensation and is the seat of intelligence
- 427-347BCE - *Plato* - teaches at Athens. Believes brain is seat of mental process
- 384-322BCE - *Aristotle* () - believes brain to be cooling agent for heart; a place in which spirits (senses) move freely. "There is nothing in the intellect that is not in the senses." (*Metaphysics*) writes about sleep; believes heart is seat of mental process
- 335-280BCE - *Herophilus* (the "Father of Anatomy"); believes ventricles are seat of human intelligence. the first scientist to systematically perform scientific dissections of human cadavers and recorded his findings in over nine works which are all lost
- 130-200AD - *Galen* - physician to gladiators, thereby gaining extensive empirical experience.

In ancient times, as the list of associations grew, it was noticed that the connections lacked rigorous specificity. In time it was learned, through repeated study, that injuries could be quite extensive and cause a function loss that was more widespread than the actual function of a smaller part. This continues to be a problem today, leading to other techniques for acquiring information.

stimulation TIMELINE
- 1876 - David Ferrier Employing carefully controlled ablation procedures and faradic stimulation of the brain, Ferrier succeeded in mapping sensory and motor areas of the cortex across a wide range of species. Ferrier's work served to confirm the establishment of sensorimotor analysis as the dominant explanatory paradigm within psychology. Publishes first detailed map of the cerebral motor cortex. [FERRIER 1876]. Clinical applications of cortical localization [FERRIER 1886].
- 1884 - 1886 - Victor Horsley performs the first operation on the human brain employed CSM (Cortical stimulation mapping) to further grasp the structure and function of the

pre-Rolandic and post-Rolandic areas, also known as the pre central gyrus and post central gyrus. Prior to the development of more advanced methods. Victor Horsley's studies on motor response to faradic electrical stimulation of the cerebral cortex, internal capsule and spinal cord became classics of the field. These studies were later translated to his pioneering work in the neurosurgery for epilepsy.
- early 1900s - Charles Sherrington determines difference between motor cortex and the sensory cortex.
- 1904 Sherrington, delivered the Silliman lectures at Yale University
- 1906 Published a compendium, expressing his theory that the nervous system acts as the coordinator of various parts of the body and that the reflexes are the simplest expressions of the interactive action of the nervous system, enabling the entire body to function toward one definite end at a time. (SHERRINGTON 1906)
- 1937 and 1938 delivered the Gifford lectures at the University of Edinburgh - focused on J.Fernel and his times explored philosophical thoughts about the mind, the human existence, and God, in connection with natural theology.
- 1940, revised 1951 published (SHERRINGTON 1940, 1951) as *Man on His Nature..*
- early 1900s - Harvey Cushing expanded studies and moved CSM from an experimental technique to one that became a staple neurosurgery technique using cortical stimulation mapping to treat epilepsy.increasing the neurosurgeons effectiveness in utilizing a more updated picture of the brain.
- 1937 - Wilder Penfield, before operating, stimulated the brain with electrical probes while the patients were conscious on the operating table (under only local anesthesia), and observed their responses. In this way he could more accurately target the areas of the brain responsible, reducing the side-effects of the surgery. shows that stimulating the precentral gyrus elicited a response contralaterally; a significant finding given that it correlated to the anatomy based on which part of the brain was stimulated.

association - Another learning technique was used by others who did not have access to injured organisms, but did provide care to individuals who suffered maladies not due to injuries. Physicians, such as Hippocrates, saw so many illnesses that he approached in a characteristic way, called today Natural History of Disease, a way he taught to others, a way so successful in making diagnoses and still used today, earning him the title of *Father of Western Medicine*. Apparently the technique worked in the following way: Having seen a number of people who had similar histories, he formulated commonalities that categorized the malfunction, and presumed that the problem lay in that body part.

By careful cross-categorizing he was able to distinguish symptoms that were common to many illnesses (such as fever) from organ specific conditions (such as movement disorders) and further discriminate those due to arthritis from those due to limb injury.

This method also did not have a high degree of specificity to identify brain parts, but it did advance understanding beyond the common understanding of the time[14].

Much of this technique was combined with the earlier system of associating symptoms with demonstrated pathologic changes and the autopsy became a valuable teaching experience for all physicians. Gradually, imaging techniques became more sophisticated, and provided the same information during life, improving both the provision of care and the advancement of knowledge.

reflection - A third learning technique developed when the first two groups built up a body of knowledge extensive enough to allow thinkers to draw data as needed when a structure in question either stood physically between two structures of known function or stood figuratively between two functions with presumed association. This technique became particularly useful when most of the obvious structures had been identified and the connections between them were found to be responsible for the critical changes[15] in performance.

Brain imaging techniques [Demitri] allow viewing the human brain, without invasive neurosurgery. Electromagnetic radiation is a spectrum that includes radio, microwave, light, and ionizing radiation. X-rays are at the lower end of the ionizing range, but do have power to disrupt atoms in the body, and needs to be used judiciously

---

[14] Egyptians discarded the brain on mummification; apparently seeing no purpose for it in the underworld; Aristotle considered the brain to be a cooling system for the heart - though he did recognize the intellect as a function of the brain.

[15] Remember, most of the space between neurons is occupied by axons and dendrites, which are extensions of the neurons, and carry signals that can be neutral, stimulatory or inhibitory. The number of connections in these networks easily equals the number of stars in the milky way. Though there are now micro and even nano instruments to study the individual cells; to study each one is like trying to understand the function of a city by talking to the individual citizens.

Brain imaging techniques, summarized

US	High-frequency sound waves reflect off body structures	reflected waves form picture on computer	no ionizing radiation exposure
X-ray	electromagnetic radiation passed through the body	relative penetration recorded on a film.	ionizing radiation exposure
CT	x-ray source circles the patient, its beam after passing through is sampled by one of the many detectors that line the machine's circumference.	reveals gross brain features based on differential absorption of X-rays. Structure not resolved well.	ionizing radiation exposure
MRI	Machine similar to the CT, but a strong magnetic field excites the Hydrogen atoms in tissue.	measures the radio frequencies emitted by Hydrogen atoms.	no ionizing radiation
MRA	Same as MRI but magnetic field oscillation and the resonance frequency measurement is modified		blood vessel flow visibility enhanced.
fMRI	responses to neural activity measured by detecting the changes in blood oxygenation and blood flow.	↑oxygen use and ↑ blood flow seen as a more active brain area.	can map brain parts involved in a particular mental process.
PET	short-lived radioactive material used to map functional processes in the brain.	material undergoes radioactive decay, emitting positron	High radioactivity areas associated with brain activity.
EEG	non-invasive measurement of the electrical activity of the brain by recording from electrodes placed on the scalp	resulting traces from electrical signal of a large number of neurons.	capable of detecting electrical activity changes on a millisecond-level
MEG	measure the magnetic fields produced by brain's electrical activity via extremely sensitive devices	localize pathology, determine function, neurofeedback.	
NIRS	Shining near infrared light through the skull and detecting how much the remerging light is attenuated.	optical technique for measuring blood oxygenation in the brain	provide an indirect measure of brain activity. Saturday, May 17, 2014

KEY:
- U/S   Ultra-sound
- X   X-ray
- CT   Computerized Tomography
- MRI   Magnetic Resonance Imaging
- MRA   Magnetic Resonance Arteriography
- fMRI   functional Magnetic Resonance Imaging
- PET   Positron Emission Tomography
- EEG   Electroencephalography
- MEG   Magnetoencephalography
- NIRS   Near infrared spectroscopy

GLOSSARY, Annotated

Two kinds of reality walked into a bar; one visible, the other not visible.
The bartender said to the one he could see, "What will you have?"
The visible reality asked for two beers.
Bartender says he'd draw one, then when it's done, he'd draw another.
"You don't understand," said the visible reality, "The other's for my invisible brother."
"Gotta explain that to me," said the bartender, "never heard of visible and invisible reality."
The visible entity heaved a sigh, and a tender draught came from along side him.
He explained:

Throughout history there have been two kinds of reality; one visible, the other invisible. The first evidence we have of invisible reality was in connection with folklore and religion. Prior to that time, it is reasonable to suppose that mankind dealt with the visible reality, just as did the rest of the animal kingdom; and elaborated the tales, religion, and philosophy as a consequence of literacy and increased brain capacity [WOLF 2007], [STEVKO 2014]. Whether or not this ability is happening in other animals is not known, although some think it may be. Some even posit evidence for group communication and defense amongst plants [POLLAN 2013].

As classical western philosophy flourished in Greece through the pre-Enlightenment, dualism stayed couched in differentiations like *real / ideal*, and *res extensa / res cogitans*. In the early days of modern western philosophy, a paradigm shift was attempted by speaking of:

term	meaning	knowable by
phenomena	(things as they appear)	the senses
noumena	(things as they are)	not the senses
		nous (pure intelligence)

This shift accomplished two things -- for one, it relieved the discomfort that ancient philosophy had with things in the tangible world and their deterioration. Things like ideas were safely ensconced in the "Ideal" world, but there had to be another dimension to that world to trust philosophic thinking. The world of religion also absorbed this thinking and brought their own supernatural twist to it. The second thing that it accomplished was the acceptance of the tangible world not as a concrete, unchangeable place, but as a compilation of perception based on the relationships of the atomic structures we perceived. Although not apparent at first, this fit easily into reality as the physicists came to understand it beyond that time.

The paradigm shift did create some confusion.
- things known by the senses are no longer "things as they are", but "things as they appear". The noumena (things as they are) no longer are what we see, but what used to be called Ideal. It makes sense within the paradigm, but confuses what we always accepted. If, however, we read the ancients more closely, we see that the original conception of real and ideal each had more than one dimension (see Plato's Analogy of the Divided Line, which we always rushed through to get to the Analogy of the Cave).
- *nous* translated as intelligence gives an inadequate idea of the "higher awareness", "pure intelligence", or "intuition", that is necessary for understanding. This gave rise to many notions of secret knowledge.

While these murky waters swirled in the West, the Hindu thinkers made many attempts to avoid these dualities. Some of their translations influenced Western philosophers [SCHOPENHAUER 1818,1969], who in turn magnified interest, which was again enlarged by the influence of existentialism on European populations. Many adherents of Meditation techniques, Mindfulness, and reflective thinking are finding helpful insights into non-dualistic thinking, and are able to achieve states of consciousness they describe as "higher", the awareness of awareness, and self-knowledge in the broadest sense

BIBLIOGRAPHY

Manter and Gatz's Essentials of Clinical Neuroanatomy and Neurophysiology by Sid Gilman and Sarah Winans Newman, 10th Edition, F. A. Davis Company (2002)
Adams and Victor's Principles of Neurology 10th Edition by Allan Ropper, Martin Samuels and Joshua Klein (May 30, 2014) McGraw-Hill Professional; 10 edition (2014)
Oliver Sacks - Every book by this author explores malfunctions of the nervous system and enlightens its functions.
Damasio Antonio - Every book by this author explores malfunctions of the nervous system and enlightens its functions.

citations

American Psychiatric Association. Diagnostic and Statistical Manual of Mental Disorders: DSM-IV. 4 ed. Washington, DC: American Psychiatric Association; 2000.
American Psychiatric Association. *Diagnostic and Statistical Manual of Mental Disorders* (Fifth edition). Arlington, VA: American Psychiatric Publishing. (2013)
Archontaki, D., Lewis, G. J. and Bates, T. C. (2012). Genetic influences on psychological well-being: A nationally representative twin study. *Journal of Personality*, **81**. 221-230.
Brodmann, K., Vogt, C., Vogt, O. *architectonic results*, annual meeting, The German Psychiatric Society, Jena. (1903)
Brodmann, K., *Vergleichende Lokalisationslehre der Grosshirnrinde,* Verlag von Johann Ambrosius Barth (1909); *Localisation in the Cerebral Cortex, The Principles of Comparative Localisation in the Cerebral Cortex Based on Cytoarchitectonics,* Springer (2006)
Cajal SR. 1894. La fine structure des centres nerveux. *Proc R Soc Lond* **55**: 444–468.
Centers for Disease Control and Prevention (CDC), *Pediatrics* March 1994.
Chomsky, Noam, *Syntactic Structure,* Mouton & Co. (1957)
Corr, Philip J.; Perkins, Adam M. (2006). "The role of theory in the psychophysiology of personality: From Ivan Pavlov to Jeffrey Gray". *International Journal of Psychophysiology* 62 (3): 367–376.
Darwin, C., *The Descent of Man and Selection in Relation to Sex* (1871)
Damasio A.R. (1994). *Descartes' Error: Emotion, Reason, and the Human Brain.* (2nd ed.:2005)
Damasio, H.; Grabowski, T.; Frank, R.; Galaburda, A.M.; Damasio, A.R. (1994). *The return of Phineas Gage: Clues about the brain from the skull of a famous patient. Science* **264** (5162): 1102–1105.
Demitri, M. . *Types of Brain Imaging Techniques.* Psych Central. (2007)
Empedocles (490-430BC), *Empedocles: The Extant Fragments*
Ferrier, David, *The Functions of the Brain,* (1876)
Ferrier, David,*The Localization of Brain Disease,* (1886)
Fitzgerald DA, Isaacs D., *Genotype-phenotype correlations with personality traits of healthcare professionals: a new use for the Human Genome Project.* Med J Aust. 2002 Apr 1;176(7):339-40.
Gall, F. J., *On the Functions of the Brain and of Each of Its parts: With Observations on the Possibility of Determining the Instincts, Propensities, and Talents, Or the Moral and Intellectual Dispositions of Men and Animals, by the Configuration of the Brain and Head,* Volume 1. Marsh, Capen & Lyon. (1835)
Harper, Douglas, *Online Etymology Dictionary* (© 2001-2014)
Johnson, C. P.; Myers, S. M. "Identification and Evaluation of Children With Autism Spectrum Disorders". *Pediatrics* **120** (5): 1183–1215. (2007).
Johnson W. E., British logician (1858-1931), neologism in Passmore, John, *A Hundred Years of Philosophy* (2nd ed.). London: Penguin (1957)
Kolb & Whishaw: *Fundamentals of Human Neuropsychology* (2003)
LeDoux JE (2000) Emotion circuits in the brain. Annu Rev Neurosci. 23, 155-184.
LeDoux JE (2002) Emotion, Memory, and the Brain. Sci Am 12:62-71.
Lehrer, J., *Proust was a Neuroscientist* Houghton Mifflin Co., New York 2007
Marden, P.M., Smith, D.W., and McDonald, M.J.: Congenital anomalies in the newborn infant, including minor variations. J. Pediatr. 64:357, 1964
Merriam-Webster's UnabridgedDictionary, version 3.0, 2003

Mesulam, M.-M *Principles of Behavioral Neurology (Contemporary Neurology, No 26)* by (1985), F.A, Davis.
Meynert, T. *L'amentia*. Paris: Presses Universitaires de France. (1890)
OED, Oxford University Press (2013).
Penfield, W., and Erickson,T., C *Epilepsy and Cerebral Localization: A Study of the Mechanism, Treatment and Prevention of Epileptic Seizures...* Charles C Thomas, 1941.
Penfield, Wilder. *The Mystery of the Mind : A Critical Study of Consciousness and the Human Brain.* Princeton University Press, 1975
Pollan, Michael, *THE INTELLIGENT PLANT,* The New Yorker. December 23, 2013
Ramachandran, V.S.; Hirstein, William (1999). "The Science of Art: A Neurological Theory of Aesthetic Experience". *Journal of Consciousness Studies* **6** (6-7): 15–51.
Random House Dictionary, © Random House, Inc. (2014).
Ranson, S.W., *Anatomy of the Nervous System*, W. B. Saunders, 1920. includes the human cerebral cortex as delineated by Korvinian Brodmann on the basis of cytoarchitecture.
Reynolds, Cecil R.; Goldstein, Sam (1999). *Handbook of neurodevelopmental and genetic disorders in children*. New York: The Guilford Press. pp. 3–8.
Shakespeare, *Julius Caesar*, act 4, scene 3
Sherrington,C., *The Integrative Action of the Nervous System*, (1906)
Sherrington,C,, *Man on His Nature*. (1940, rev. 1951)
Schopenhauer, A., *The World as Will and Representation* (1818), Dover (1966)
Smith D,W., *Recognizable Patterns of Human Malformation*, Saunders, 1976.
Stevko, R.M., Balsley, M., Segar, W.E., *Primary polydipsia—compulsive water drinking:*
Report of two cases, The Journal of Pediatrics, Volume 73, Issue 6 , Pages 845-851, (1968)
Stevko, R. *Shades of Meaning*, Graven Image Publishers. 2013
Stevko, R. *Before Philosophy*, Graven Image Publishers. 2014
Wolf, Maryanne, *Proust and the Squid*, HarperCollins Publishers, 2007
Wittgenstein, L., *Tractatus Logico-Philosophicus* (1921)
Zeki, Semir. *Inner Vision: an exploration of art and the brain*. Oxford University Press, (1999).
Zeman A., *Consciousness*, Brain. 2001 Jul;124(Pt 7):1263-89.

www.ingramcontent.com/pod-product-compliance
Lightning Source LLC
Chambersburg PA
CBHW080831170526
45158CB00009B/2549